AF360809

Advances in Bio-Inspired Robots

Advances in Bio-Inspired Robots

Editors

TaeWon Seo
Dongwon Yun
Gwang-Pil Jung

MDPI • Basel • Beijing • Wuhan • Barcelona • Belgrade • Manchester • Tokyo • Cluj • Tianjin

Editors
TaeWon Seo
School of Mechanical
Engineering, Hanyang
University
Korea

Dongwon Yun
Department of Robotics
Engineering, Daegu Gyeongbuk
Institute of Science &
Technology (DIGST)
Korea

Gwang-Pil Jung
Department of Mechanical &
Automotive Engineering,
SeoulTech
Korea

Editorial Office
MDPI
St. Alban-Anlage 66
4052 Basel, Switzerland

This is a reprint of articles from the Special Issue published online in the open access journal *Applied Sciences* (ISSN 2076-3417) (available at: https://www.mdpi.com/journal/applsci/special_issues/bio-inspired_robots).

For citation purposes, cite each article independently as indicated on the article page online and as indicated below:

LastName, A.A.; LastName, B.B.; LastName, C.C. Article Title. *Journal Name* **Year**, *Volume Number*, Page Range.

ISBN 978-3-0365-2512-9 (Hbk)
ISBN 978-3-0365-2513-6 (PDF)

Contents

About the Editors

TaeWon Seo received his B.S. and Ph.D. degrees from the School of Mechanical and Aerospace Engineering, Seoul Nat'l Univ., Korea in 2003 and 2008, respectively. He is an Associate Professor at the School of Mechanical Engineering, Hanyang Univ., Korea. Before Hanyang Univ., he was a postdoctoral researcher at Nanorobotics Lab., Carnegie Mellon Univ., a visiting professor at Biomimetic Millisystems Lab., UC Berkeley, visiting scholar at University of Michigan, and an Associate Professor at the School of Mechanical Engineering, Yeungnam Univ., Korea. His research interests include robot design, analysis, control, optimization, and planning. Dr. Seo received the Best Paper Award of the IEEE/ASME Transaction on Mechatronics in 2014. He was a Technical/Associate Editor of IEEE/ASME Transaction on Mechatronics and Intelligent Service Robots, and was Associate Editor of IEEE Robotics and Automation Letters.

Dongwon Yun received his B.S. degree in mechanical engineering from Pusan National University, Korea, in 2002, an M.S. degree in mechatronics engineering in 2004 from GIST, Korea, and a Ph.D. degree in mechanical engineering from KAIST, Korea, in 2013, respectively. He was a Senior Researcher for Korea Institute of Machinery and Materials from 2005 to 2016. He joined the Department of Robotics Engineering, DGIST in 2016 and he is an Associate Professor in DGIST. His research interests include bio-mimetic robot systems, industrial robot systems and mechatronics, soft robotics, and sensors and actuators. He was a board member of Korean Society of Mechanical Engineering(KSME) and Military Robotics Society(MRS), and a Senior member of IEEE. He was also on an organizing committee for the International Conference on Multisensor Fusion and Integration for Intelligent Systems (MFI 2019) and International Conference of Mechatronics Technology (ICMT).

Gwang-Pil Jung received his B.S. degree in mechanical engineering from the KAIST, Daejeon, Korea, in 2010 and the Ph.D. degree in mechanical engineering from Seoul National University, Seoul, Korea. He is an Assistant Professor at the Department of Mechanical and Automotive Engineering, SeoulTech, Korea. His current research interests include the design and fabrication of biologically inspired robots and novel mechanisms using smart materials, structures, and actuators. Dr. Jung received the Best Video Award in IROS 2013. He was an Associate Editor of Journal of Mechanical Science and Technology.

Preface to "Advances in Bio-Inspired Robots"

Bio-inspiration is a good starting point for designing innovative mechanical systems, including robots. Additionally, it is a good way to solve engineering design and control problems. Recently, creative design has become very important in the robotics field, not just following previous design solutions but also suggesting novel system designs for various tasks. Bio-inspiration is becoming more and more popular in order to obtain innovative solutions inspired by animals and insects.

In this Special Issue, nine excellent papers are published on advances in bio-inspired robots. The papers are categorized into three groups as follows:

A. Biomimetic robot design;
B. Mechanical system design from bio-inspiration;
C. Bio-inspired analysis on a mechanical system.

TaeWon Seo, Dongwon Yun, Gwang-Pil Jung
Editors

Editorial

Special Issue on Advances in Bio-Inspired Robots

TaeWon Seo [1,*], Dongwon Yun [2] and Gwang-Pil Jung [3]

[1] School of Mechanical Engineering, Hanyang University, Seoul 04763, Korea

[2] Department of Robotics Engineering, Daegu Gyeongbuk Institute of Science & Technology (DIGST), Daegu 42988, Korea; mech@dgist.ac.kr

[3] Department of Mechanical & Automotive Engineering, SeoulTech, Seoul 01811, Korea; gpjung@seoultech.ac.kr

* Correspondence: taewonseo@hanyang.ac.kr

Bio-inspiration is a good starting point of designing innovative mechanical systems, including robots. Additionally, it is a good way to solve engineering design and control problems. Recently, creative design has become very important in the robotics field, not just following previous design solutions but also suggesting novel system designs for various tasks. Bio-inspiration is becoming more and more popular in order to obtain innovative solutions inspired by animals and insects.

In this Special Issue, nine excellent papers are published on advances in bio-inspired robots. The papers are categorized into three groups as follows:

A. Biomimetic robot design
B. Mechanical system design from bio-inspiration
C. Bio-inspired analysis on a mechanical system

The detail contents can be summarized as follows.

1. Biomimetic Robot Design

1.1. Soft Jumping Robot Using Soft Morphing and the Yield Point of Magnetic Force

Soft-morphing, deformation control by fabric structures and soft-jumping mechanisms using magnetic yield points are studied. The soft jumping mechanism can transfer energy more efficiently and stably using an energy storage and release mechanism and the rounded ankle structure designed using soft morphing [1].

1.2. Snake Robot with Driving Assistant Mechanism

A driving assistant mechanism is proposed for a snake robot, which assists locomotion without additional driving algorithms and sensors. The driving assistant mechanism can prevent roll down on a slope and can increase the locomotion speed [2].

1.3. A Miniature Flapping Mechanism Using an Origami-Based Spherical Six-Bar Pattern

A novel transmission is proposed for DC motor-based flapping-wing micro aerial vehicles. The proposed origami-based fabrication method reduces the number of relative moving components by replacing rigid links and pin joints with facets and folding joints, which shortens the assembly process and reduces friction between components [3].

2. Mechanical System Design from Bio-Inspiration

2.1. Bioinspired Divide-and-Conquer Design Methodology for a Multifunctional Contour of a Curved Lever

Bioinspired design methodology for a multifunctional lever is proposed based on the morphological principle of the lever mechanism in the Salvia pratensis flower. Four partial contours are designed to satisfy three types of functional requirements. The final design for the lever contour is manufactured and verified with visual measurement experiments [4].

1

2.2. Cable Tension Analysis Oriented the Enhanced Stiffness of a 3-DOF Joint Module of a Modular Cable-Driven Human-Like Robotic Arm

Inspired by the structure of human arms, a modular cable-driven human-like robotic arm is developed for safe human–robot interaction. Due to the unilateral driving properties of the cables, the robotic arm is redundantly actuated and its stiffness can be adjusted by regulating the cable tensions [5].

2.3. A Novel Type of Wall-Climbing Robot with a Gear Transmission System Arm and Adhere Mechanism Inspired by Cicada and Gecko

A novel type of wall-climbing robot is proposed with a new gear transmission system arm and an adherence mechanism inspired by cicadas and geckos. The adherence force experiments demonstrate that the bionic spines and bionic materials achieved good climbing on cloth, stone, and glass surfaces [6].

3. Bio-Inspired Analysis on A Mechanical System

3.1. Control Strategy for Direct Teaching of Non-Mechanical Remote Center Motion of Surgical Assistant Robot with Force/Torque Sensor

A control strategy is proposed from bio-inspiration that secures both the precision and manipulation sensitivity of remote center motion with direct teaching for a surgical assistant robot. Instead of the bulky mechanically constrained remote center motion mechanism, a conventional collaborative robot is used to mimic the wrist movement of a scrub nurse [7].

3.2. Empirical Modeling of Two-Degree-of-Freedom Azimuth Underwater Thruster Using a Signal Compression Method

Empirical modeling of a two-degree-of-freedom (DoF) azimuth thruster is presented based on bio-inspiration using the signal compression method. Empirical models of force and moment for rotational motion were derived for practical use through frequency analysis [8].

3.3. Energy-Efficient Hip Joint Offsets in Humanoid Robot via Taguchi Method and Bio-Inspired Analysis

The offsets of hip joints in humanoid robots are optimized via the Taguchi method to maximize energy efficiency. Through two optimization stages, near-optimal results are obtained for small power consumption [9].

Acknowledgments: This Special Issue would not have been possible without the help of a variety of talented authors, professional reviewers, and the dedicated editorial team of *Applied Sciences*. Thank you to all the authors and reviewers for this opportunity. Finally, thanks to the *Applied Sciences* editorial team.

Conflicts of Interest: The author declares no conflict of interest.

References

1. Jeon, G.-H.; Park, Y.-J. Soft Jumping Robot Using Soft Morphing and the Yield Point of Magnetic Force. *Appl. Sci.* **2021**, *11*, 5891. [CrossRef]
2. Kim, J.; Moon, J.; Ryu, J.; Lee, G. Bioinspired Divide-and-Conquer Design Methodology for a Multifunctional Contour of a Curved Lever. *Appl. Sci.* **2021**, *11*, 6015. [CrossRef]
3. Kim, M.; Zhang, Y.; Jin, S. Control Strategy for Direct Teaching of Non-Mechanical Remote Center Motion of Surgical Assistant Robot with Force/Torque Sensor. *Appl. Sci.* **2021**, *11*, 4279. [CrossRef]
4. Bian, S.; Xu, F.; Wei, Y.; Kong, D. A Novel Type of Wall-Climbing Robot with a Gear Transmission System Arm and Adhere Mechanism Inspired by Cicada and Gecko. *Appl. Sci.* **2021**, *11*, 4137. [CrossRef]
5. Jeong, C.-S.; Kim, G.; Lee, I.; Jin, S. Empirical Modeling of 2-Degree-of-Freedom Azimuth Underwater Thruster Using a Signal Compression Method. *Appl. Sci.* **2021**, *11*, 3517. [CrossRef]
6. Bae, S.-Y.; Koh, J.-S.; Jung, G.-P. A Miniature Flapping Mechanism Using an Origami-Based Spherical Six-Bar Pattern. *Appl. Sci.* **2021**, *11*, 1515. [CrossRef]

7. Yang, K.; Yang, G.; Zhang, C.; Chen, C.; Zheng, T.; Cui, Y.; Chen, T. Cable Tension Analysis Oriented the Enhanced Stiffness of a 3-DOF Joint Module of a Modular Cable-Driven Human-Like Robotic Arm. *Appl. Sci.* **2020**, *10*, 8871. [CrossRef]
8. Bae, J.; Kim, M.; Song, B.; Jin, M.; Yun, D. Snake Robot with Driving Assistant Mechanism. *Appl. Sci.* **2020**, *10*, 7478. [CrossRef]
9. Kim, J.; Yang, J.; Yang, S.T.; Oh, Y.; Lee, G. Energy-Efficient Hip Joint Offsets in Humanoid Robot via Taguchi Method and Bio-inspired Analysis. *Appl. Sci.* **2020**, *10*, 7287. [CrossRef]

applied sciences

MDPI

Article

Soft Jumping Robot Using Soft Morphing and the Yield Point of Magnetic Force

Gang-Hyun Jeon and Yong-Jai Park *

Robot & Mechanism LAB, Biohealth-Machinery Convergence Engineering, Kangwon National University, Chuncheon 24341, Gangwon, Korea; wjsrkdgus1994@gmail.com
* Correspondence: yjpark@kangwon.ac.kr; Tel.: +82-33-250-6371

Abstract: In this paper, soft-morphing, deformation control by fabric structures and soft-jumping mechanisms using magnetic yield points are studied. The durability and adaptability of existing rigid-base jumping mechanisms are improved by a soft-morphing process that employs the residual stress of a polymer. Although rigid body-based jumping mechanisms are used, they are driven by multiple components and complex structures. Therefore, they have drawbacks in terms of shock durability and fatigue accumulation. To improve these problems, soft-jumping mechanisms are designed using soft polymer materials and soft-morphing techniques with excellent shock resistance and environmental adaptability. To this end, a soft jumping mechanism is designed to store energy using the air pressure inside the structure, and the thickness of the polymer layer is adjusted based on the method applied for controlling the polymer freedom and residual stress deformation. The soft jumping mechanism can transfer energy more efficiently and stably using an energy storage and release mechanism and the rounded ankle structure designed using soft morphing. Therefore, the soft morphing and mechanisms of energy retention and release were applied to fabricate a soft robot prototype that can move in the desired direction and jump; the performance experiment was carried out.

Keywords: soft robot; soft jumping robot; soft morphing; residual stress; magnetic yield point

check for
updates

Citation: Jeon, G.-H.; Park, Y.-J. Soft Jumping Robot Using Soft Morphing and the Yield Point of Magnetic Force. *Appl. Sci.* **2021**, *11*, 5891. https://doi.org/10.3390/app11135891

Academic Editor: Alessandro Gasparetto

Received: 31 May 2021
Accepted: 23 June 2021
Published: 24 June 2021

Publisher's Note: MDPI stays neutral with regard to jurisdictional claims in published maps and institutional affiliations.

1. Introduction

The development of Fourth Industrial Revolution technologies has resulted in the extensive application of these technologies in various fields and industries; various robots and production processes have been recently developed for more efficient and safe work. Robots produced through new materials and new manufacturing methods are applied to perform various tasks that were performed manually. However, with an increase in the number of robots replacing human intervention in various fields and environments, the problem of robot control and driving in various environments is gaining considerable research attention. Robots made up of rigid bodies are designed and controlled according to a particular process or environment, and therefore, there is a problem in that the design or control of the drive part must be changed when the environment of use changes. High-tech and high-performance robots are being developed, such as Cheetah [1] at MIT and Big Dog [2] at Boston Dynamics to develop a robot that can be applied immediately to multiple environments without changing the design or control of the drive. However, these robots have limitations in terms of accessibility, manufacturing process, and the high cost of the technology required to build the robot. Therefore, a new type of robot is actively being developed to improve the adaptability and durability of robots and to build highly accessible robots that can be efficiently applied to various environments without changing the design or control of the drives. Soft robots are made of soft materials, such as polymers, that absorb shocks well and are adaptable to objects. Further, these materials have advantages in that they can be used to build robots that can be applied to various environments with simpler control. However, it is difficult to apply a drive that requires instantaneous release of great force, such as a jump, to a soft robot.

To solve these problems and further highlight the advantages of soft robots, various soft robots and driving parts are being developed that can be integrated while preserving the characteristics and advantages of soft robots. The related studies include driving methods that employ properties of substances such as SMA (shape memory alloy) and SMP (shape memory polymer) [3–7] that deform in response to specific conditions such as temperature and light, soft robots [8–10] that use instantaneous energy generated by the explosion of gas compounds, and robots [11,12] that use mechanical designs [13–17]. Among biomimetic robots [18–24], soft morphing [25–27], which creates residual stresses in structures or inserts actuators to change shape [28,29], and robots that use soft materials and pneumatic drive [30] can be applied without significantly degrading their performance.

In this paper, the thickness of the polymer layer is adjusted by the needed shape. Further, a one-way tensile force is applied in the soft morphing process for stacking different kinds of polymers and forming residual stresses on the polymer. The non-stretching fabric material is inserted in layers to control the deformation and shape of polymers with high degrees of freedom to suit their needs. In addition, we produced prototypes of soft robots that could overcome various environments such as crawling and jumping using soft morphing, magnetic yield, and air chambers inserted inside the structure through the storage and release of energy implemented by the soft robot. The final goal of this research is to install an air pressure drive system inside a soft jumping robot and enable the robot to perform tasks such as investigating and transmitting light objects on behalf of humans.

2. Fabrication of Polymer Structures and Formation of Residual Stresses

2.1. Soft Morphing

Soft morphing structures are mostly made of soft materials with no restrictions on the degrees of freedom, and they are characterized by continuous and soft movement. By using the continuous soft movement of soft morphing, one can not only achieve natural and flexible movements that cannot be realized by robots with rigid bodies, but it can also help overcome limitations of rigid materials that are vulnerable to impact. In the case of polymers, even if external shocks such as shock or distortion are applied, they absorb shock smoothly and do not break. Furthermore, the hardness of the soft material forming the structure can be adjusted to a desired degree of rigidity. Therefore, various studies on soft morphing [27–29] have been actively conducted owing to these advantages.

2.2. Residual Stresses

Residual stress is defined as the energy applied to an object that remains indefinitely in the object while retaining the strength and direction of the residual stress. Because of residual stress, the object can have various forms and numbers of stable states. In this paper, the residual stress is induced in a polymer structure to minimize the driving part by steady state deformed by residual stress and produce a soft mechanism that highlights the advantages of soft morphing. Through residual stress induced in a polymer structure the robot's shape and jumping motion can be completed and a part of the recovery process can be achieved. To produce a ductile polymer structure, Cubicon's Style NEO-A22 model and PLA+ 3D printer filament was used to produce 100% molds of interior filling (0.014 m ×0.003 m). A suitable amount of smooth-on soft polymer, a curing accelerator Plat Cat, and Ecoflex0050 were mixed into the mold; depopulation was carried out at a pressure of 0.1 MPa using a vacuum depopulator for 10 min. After the deformation, the liquid polymer was poured into the mold, and an acrylic plate and a 1.5 kg weight were placed on it. Further, it was cured in an oven at 55 °C for 60 min. To make a polymer structure, the liquid polymer must be cured by mixing main material A and sub material B. The mixing ratio of the main and sub materials to fabricate polymer structures used in this paper is summarized in Table 1.

Table 1. Mixing ratio of liquid polymer.

Material	Percent (%)
Main material A	49.5
Sub material B	49.5
Cure accelerator	1

To induce residual stress inside the polymer structure using the EcoFlex0050 liquid polymer material, the elasticity of the polymer was used to stretch the polymer structure. When the polymer structure is fixed in a tensile state, the elastic stresses from inside the polymer structure maintain the stress generated by the elastic resilience and form residual stress inside. This is conducted to the maintain residual stress in a state where the structure is stretched and fixed. Other polymer structures are bonded to the upper part to interfere with the contraction of the lower polymer structures in a state where the stress caused by the elastic resilience is not eliminated. Therefore, the lower polymer structure is stretched to prevent the stress formed inside from being eliminated by the structure bonded to the upper part and the stress formed inside (Figure 1).

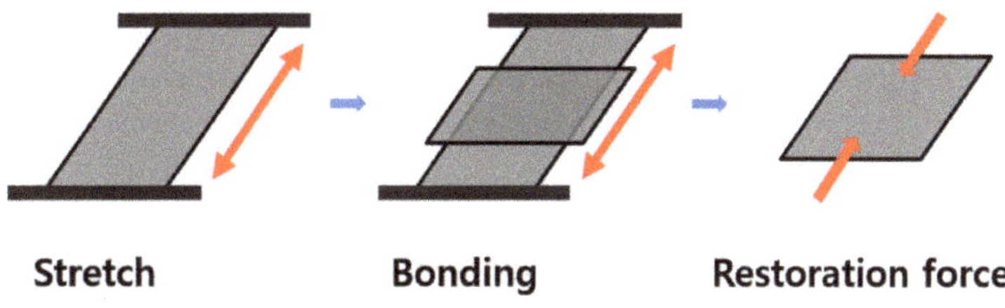

Figure 1. Process of forming residual stresses on a polymer structure.

3. Fabrication of Deformation Structures Caused by Residual Stresses

The lower polymer structure must be fixed in a tensile state until the end of the upper polymer bonding process to form a stable residual stress inside the polymer structure. A device (Figure 2) was set to hold the polymer structure in a tensile state to maintain the tensile state of the lower polymer structure throughout the work process.

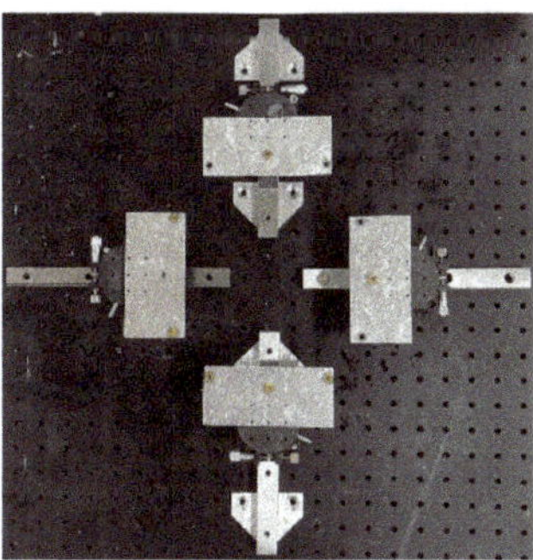

Figure 2. Process of forming residual stresses on a polymer structure.

First, a perforated metal plate was set to hold the polymer structure securely to the floor without shaking. Next, we fix the polymer structure 0.005 m away from both ends of the EcoFlex0050 and it is stretched and secured at 0.02 m. The tensile and fixed polymer structures prevent the stress caused by the elastic resilience from being eliminated in the manufacturing process and allow the stable manufacture of upper polymer structures. When the lower polymer structure is stretched and fixed, as shown in Figure 3a, the mold of upper polymer structure is attached to the lower polymer structure. The flat plate is laid in the empty space under the polymer structure so that work can be conducted more accurately. Considering the Poisson ratio of the polymer structure so that the polymer structure can be deformed uniformly, the center of the mold and the center of the polymer structure are bonded in the same way, as shown in Figure 3b.

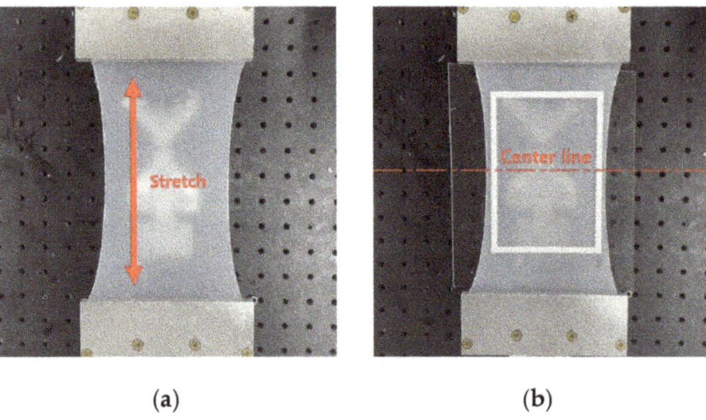

(a) (b)

Figure 3. Process of forming residual stresses on a polymer structure. (**a**) Stretch-fixed lower polymer structure; (**b**) working position of upper polymer structure.

Dragon Skin Fx pro liquid polymer, which has low viscosity and higher hardness compared to that of EcoFlex0050, is used to resist shrinkage. To remove bubbles, a 367.75 W vacuum chamber was used for 7 min to defoaming at a pressure of 10^5 Pa. The acrylic plate and 1.5 kg weight were raised and cured at 23 °C for 60 min to prevent the inflow of dust substances into the polymer structure and to ensure the uniform surface of the polymer structure. Finally, structures modified by residual stress as shown in Figure 4 were constructed.

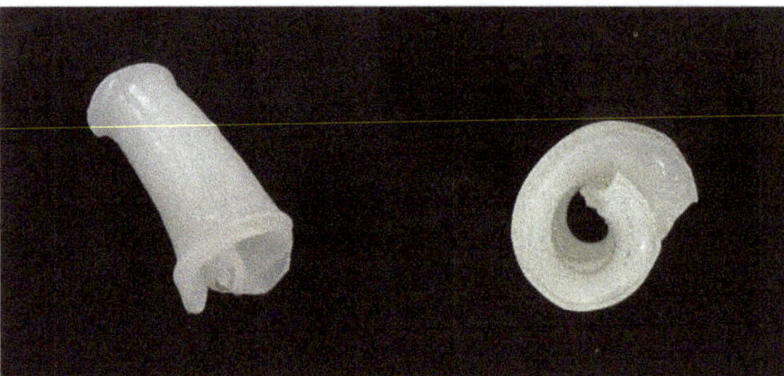

Figure 4. Structures deformed by residual stress.

4. Deformation of Soft Morphing Structures According to the Conditions of Lower Polymer

To determine the effect of tensile stress, elasticity, and thickness degree of the polymer structure on the deformation of the structure, the tensile property, elasticity, and thickness of lower polymer structure are changed when manufacturing the polymer structure. The vacuum defoaming time and curing time of the material and mixing ratio of the polymer structure are set to the same value and Dragon skin Fx pro was used for the upper polymer structure. The properties of the material used in the experiment are shown in Table 2.

Decreasing the tensile force 0.02 m to 0.01 m of the lower polymer structure (Figure 5a,b) causes a lower deformation of the polymer structure, which decreases the stress-energy generated inside the polymer structure. The decrease in the stress energy stored in the lower polymer structure causes a decrease in the stress asymmetry of the upper and lower polymer structures. Thus, the greater the stress asymmetry of the upper and lower polymer structures, the greater is the deformation rate of the final polymer structures, and the greater is the tensile degree of the lower polymer structures. If the elasticity of the lower polymer structure decreases by changing the material from EcoFlex0050 to EcoFlex0030 (Figure 5a,c), lower stress caused by the elastic resilience is formed inside the polymer structure during the tensile process. The deformation rate of the final polymer structure is lower because the decreased stress stores less stress internally than the existing lower polymer structure. When the tensile strength is the same as that of the raw material, the deformation rate of the final polymer structure becomes lower because the elastic restoring force decreased when the lower polymer structure is made thinner (Figure 5a,d). The experiments were carried out to verify the effect of various values of the tensile stress, elasticity, and thickness tests on the strain. When the tensile stress, elasticity, and thickness value of the lower polymer structure increases, the stress of the lower polymer also increases. Therefore, the greater the stress of the lower polymer structure is used, the greater the strain of the soft morphing structure is presented. The higher strain has the advantage of increasing a dynamic performance of a jumping robot. As a result, the 0.003 m-thick EcoFlex0050 was stretched by 0.02 m to build the soft jumping robot as the lower polymer structure. The upper polymer structure has been made by Dragon skin Fx pro.

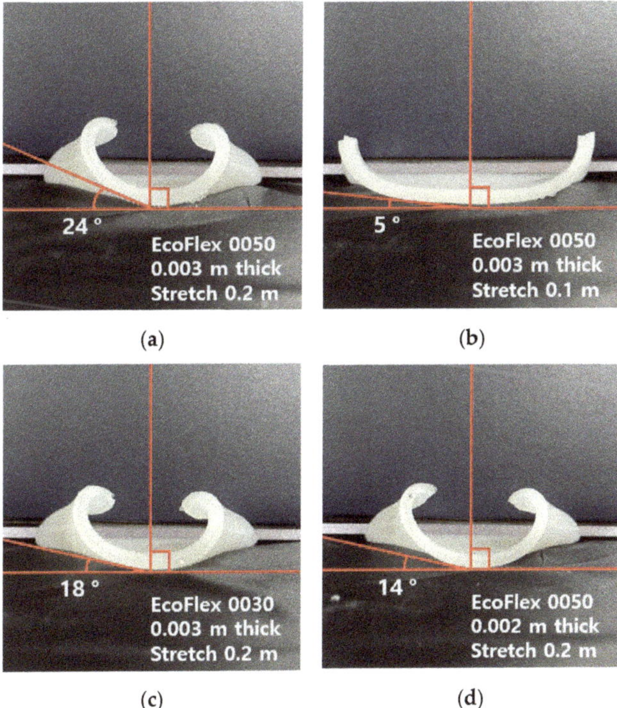

Figure 5. Differences in the strain of the polymer structures based on tensile stress. (**a**) Polymer structure made with EcoFlex0050, stretched to 0.2 m, 0.003 m thick; (**b**) polymer structure made with EcoFlex0050, stretched to 0.1 m, 0.003 m thick; (**c**) polymer structure made with EcoFlex0030, stretched to 0.2 m, 0.003 m thick; (**d**) polymer structure made with EcoFlex 0050, stretched to 0.2 m, 0.002 m thick.

Table 2. Properties of the material used in the experiment.

Properties \ Material	EcoFlex0050	EcoFlex0030	Dragon Skin Fx Pro
Shore hardness	00–50	00–30	2
Mixed viscosity	8000 cps	3000 cps	18,000 cps
Tensile strength	315 psi	200 psi	288 psi
Elongation at break	980%	900%	763%

5. Soft Jumping Mechanism Using Magnetic Force

5.1. Pneumatic Drive Fabrication

A pneumatic drive is formed inside the polymer structure to drive the polymer deformed by the residual stress formed by the tensile-bonding process. In the manufacturing process of polymer structures deformed by residual stress, the polymer is stretched to 0.02 m and the mold for the upper structure is raised on the fixed lower polymer structures. Then, some adhesives are applied to the middle of the 0.06 m × 0.02 m × 0.001 m prosthesis and the prosthesis is bonded as shown in Figure 6a to prevent liquid polymers from moving out during the manufacture of the superstructure. Therefore, the upper part of the prosthesis is in the center where the Poisson's ratio is highest, and deformation occurs the most so that the effective deformation can be obtained because of the high deformation rate. A thin layer of petrolatum is applied to the prosthesis inserted for forming the air chamber so that the air chamber can be pulled out without damage. After the upper polymer structure is completely cured, the top 0.002 m part of the prosthesis inserted into the air chamber mold is cut to remove the prosthesis from the air chambers, as in Figure 6b. The incisions of the air chamber are sealed with Dragon skin Fx pro after the pneumatic tube is inserted in the process of controlling the degrees of freedom using fabric materials.

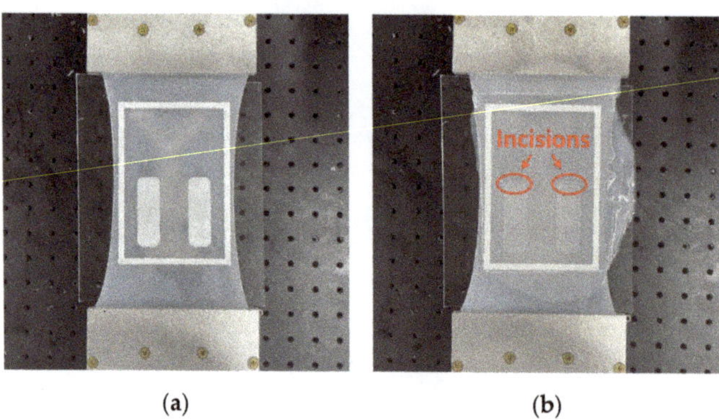

(a) (b)

Figure 6. Air chamber molding inside the polymer structure. (**a**) Mold and prosthesis work position for upper polymer structure work. (**b**) incisions for the removal of the upper polymer structure prosthesis.

5.2. Polymer Freedom Agent Using Fabric Material

In the process of manufacturing the upper polymer structure, the incision of the air chamber made by inserting the prosthesis is sutured to control the degrees of freedom of the polymer. Dragon skin Fx pro, the same material as the material used in the manufacture of the upper polymer, is coated thinly on the upper polymer structure, and a fabric material that does not extend in the upper and lower directions of the structure is placed. Then, the liquid Dragon skin Fx Pro is applied to the fabric material, which is fully absorbed by the fabric. Acrylic plates are then placed on the fabric material to ensure a smooth surface and prevent the inflow of dust substances. A 1.5 kg weight is placed on the acrylic plate so that the acrylic plate can form a smooth surface; it is secured by supporting the left and right sides to prevent slipping during the solidification process. After the solidification process, a pneumatic tube is inserted by making a 0.005 m incision at the top, as in removing the prosthesis of the air chamber. After the pneumatic tube is inserted, a Dragon skin Fx pro is applied at the joint for sealing.

The manufactured soft pneumatic drive operates in a folded and extended form depending on the presence or absence of air injection designed inside the polymer structure, as shown in Figure 7a,b.

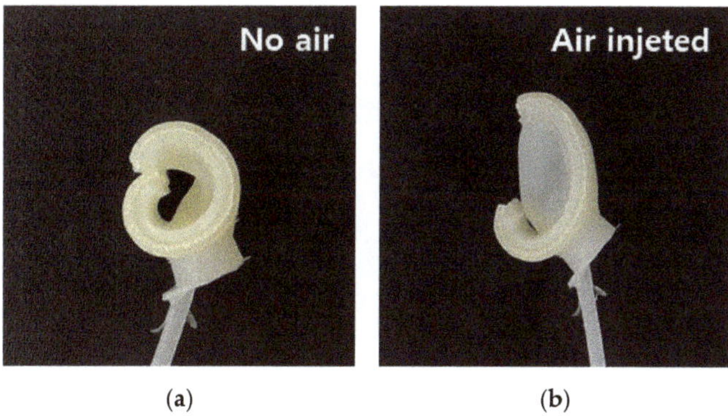

(**a**)　　　　　　　　　　　　　(**b**)

Figure 7. Movement of soft mechanisms by air injection. (**a**) State when air is not injected into the soft mechanism; (**b**) state when air is injected into the soft mechanism.

The completed composite polymer structure is shown in Figure 8. The lower polymer structure is tensile to form elastic stress. Further, the lower polymer structure is resistant to contraction, and therefore, the stress formed in the lower structure remains inside without complete resolution. In the middle of the polymer structure, there is an air chamber for the pneumatic drive. In addition, there is a fabric structure for shape control of the air chamber on top of the polymer structure.

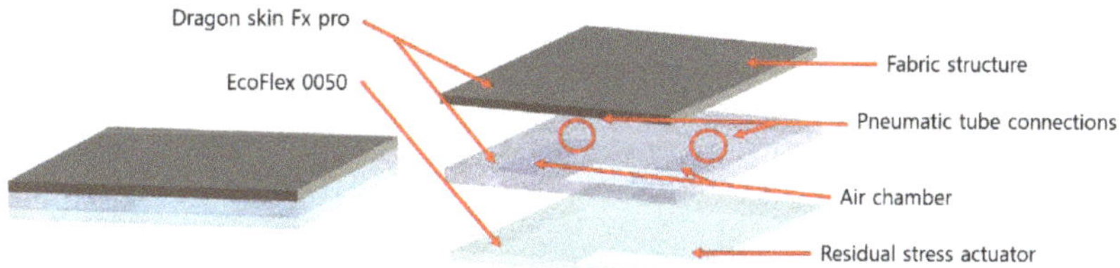

Figure 8. Decomposition of soft drive mechanisms.

5.3. Energy Storage Using Magnetism

We used permanent magnetism as a method to store energy inside a soft jumping mechanism. Permanent magnets are attached to both ends of the soft jumping mechanism as shown in Figure 9a,b.

Owing to the gravitational force of the attached permanent magnet, the air chamber cannot expand to a certain extent and produces pressure-induced energy inside. The power depends on the distance between the permanent magnets attached to both ends and it is given as:

$$F_z = -\frac{\pi \mu_0 M^2}{4} R^4 \left[\frac{1}{x^2} + \frac{1}{(x+2t)^2} - \frac{2}{(x+t)^2} \right] \tag{1}$$

The approximation can be obtained through:

$$F_z = -\frac{-3\pi\mu_0 M^2}{2} R^4 t^2 \frac{1}{x^4} \tag{2}$$

when thickness t is less than the distance x between the magnets. Equations (1) and (2) are presented and derived in [31].

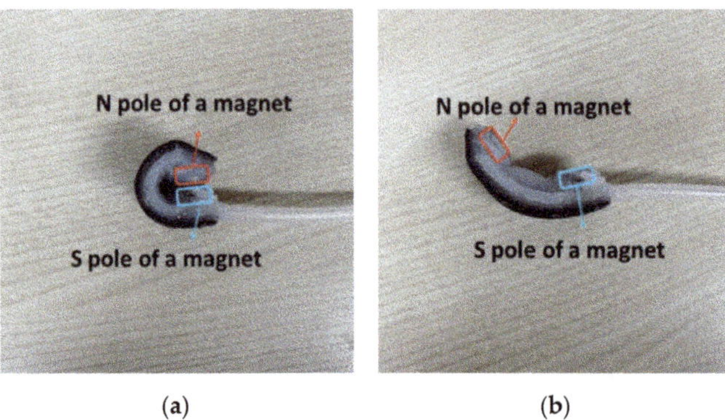

(a) (b)

Figure 9. Energy storage using permanent magnets. (**a**) No air is injected in the air chamber; (**b**) air is injected in the air chamber.

In Equations (1) and (2), F_z represents the magnet pulling force (N), π denotes the circular constant, μ_0 its permeability, M represents magnetization, R denotes the radius of the magnet, x denotes the distance between the permanent magnets, and t denotes the thickness of the permanent magnet. However, the closer the distance between permanent magnets, the less accurate are Equations (1) and (2). Thus, Equations (1) and (2) are not suitable for this study. To solve the problem, the attraction of permanent magnets corresponding to the distance was confirmed and the approximate value was obtained. A permanent magnet with a length of 0.01 m, width of 0.02 m and thickness of 0.003 m was used in the experiment; the maximum weight lifted by the magnet was measured according to the distance. The result of the experiment is shown in Figure 10.

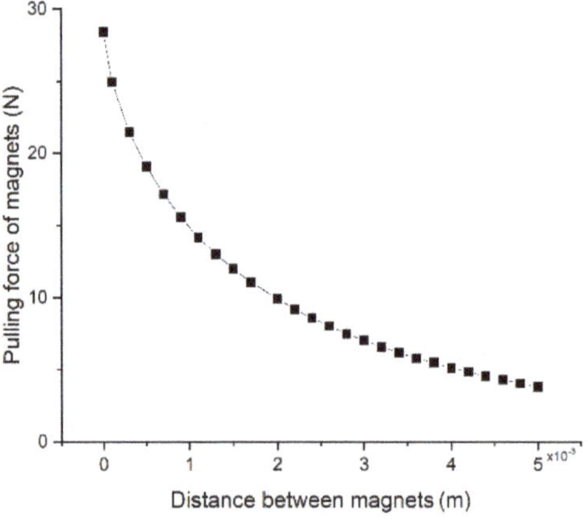

Figure 10. Pulling force of magnet according to distance.

Based on the results of the experiment, a numerical interpretation was conducted using MATLAB (MathWorks). The traction force of the permanent magnet used in the experiment can be expressed as

$$F_m = Ae^{(-Bx)} \tag{3}$$

where F_m denotes the weight that the magnet lifts; for A, it is 25.06; B; 0.44. R^2 denotes coefficient of determination of Equation (3) and it is 0.9752. Through this, the approximate energy stored in the air chamber is confirmed when using permanent magnets.

5.4. Emission of Energy Using Magnetic Yield Points

When air is injected into the air chamber inside the soft jumping mechanism, the air chamber expands. If more than a certain amount of air is injected, the permanent magnets attached to both ends prevent the expansion of the air chamber, and this increases the pressure inside the air chamber. The magnitude of the force exerted by the air chamber can be expressed as shown in:

$$F = \Delta P \times A \tag{4}$$

In Equation (4), F denotes the magnitude of the force on the air chamber, ΔP denotes amount of pressure change inside the air chamber and A denotes the cross-sectional area of the magnet pushed out by the air chamber.

The volume of the air chamber designed inside the soft jumping mechanism is length of 0.06 m, width of 0.03 m and thickness of 0.001 m. The permanent magnets (length = 0.01 m; width = 0.02 m; thickness = 0.003 m) are attached to both ends of the soft jumping mechanism. The air chamber expands until about 0.02 L of air is injected, and the magnetic field is induced when 0.035 L of air is injected as shown in Figure 11.

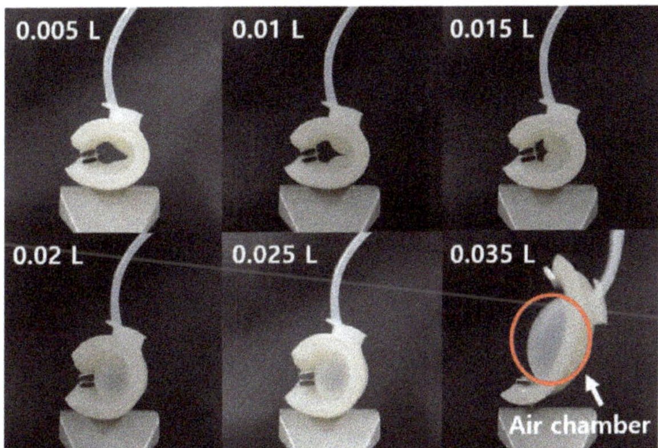

Figure 11. Pulling force of magnet according to distance.

According to Equations (3) and (4), the force of compressed air in the air chamber is 11.32 N, and the force of permanent magnets attached to soft jumping mechanism in distance 0.0015 m is 12.95 N. The actual measured distance between the permanent magnets is 0.0015 m. Although the obtained force of the air chamber was lower than that of the permanent magnet, the magnetic yield point was reached, resulting in jumping. It is because the body of the soft jumping mechanism acts as a lever at the yield point of the magnet, as shown in Figure 12. As the magnet on the upper side rotates, the actual magnetic force becomes lower compared to the calculated magnetic force.

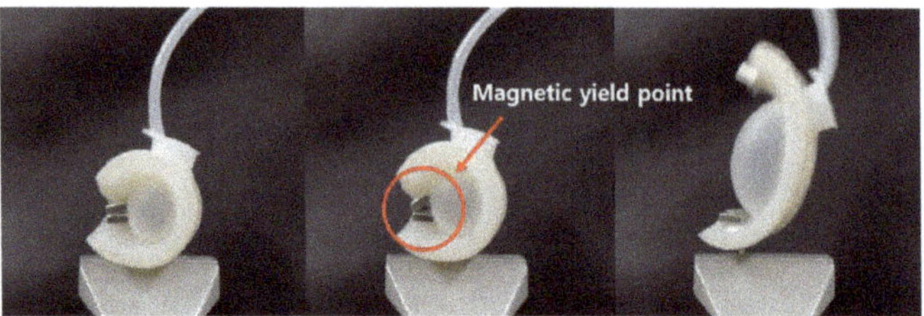

Figure 12. Movement of soft jumping mechanism at the magnetic yield point.

6. Soft Jumping Robot Fabrication Using Magnetic Surrender

6.1. Shape Regulation through Residual Stress Control by Site

Additional techniques are applied to the manufacturing stage of polymer structures deformed by residual stress to make soft jumping robots more efficient. Experiments conducted in Section 2.2 of this study confirmed that the residual stress stored in the polymer varies depending on the thickness of the polymer; i.e., the deformation rate varies depending on the thickness of the polymer. The thickness is adjusted as shown in Figure 13 so that the upper structure can adjust the residual stress inside the polymer structure to obtain the proper shape of the soft robot in the manufacturing stage of the polymer structure deformed by residual stress without additional drive design. Owing to the residual stress of the foot and body parts of the soft jumping robot, which are controlled by thickness, the soft jumping robot yields a different degree of strain.

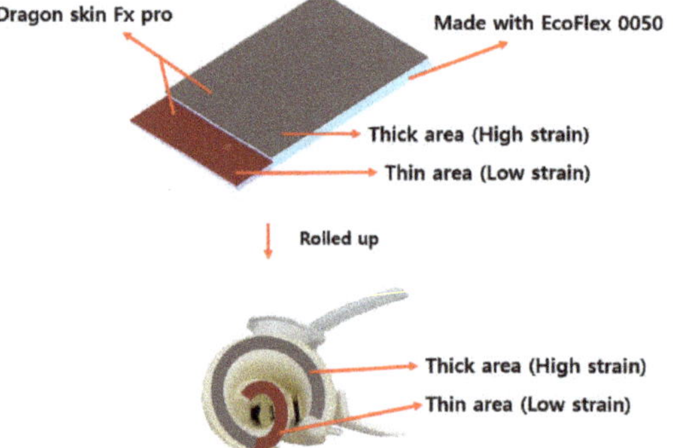

Figure 13. Residual stress control by changing the thickness of the polymer structure.

6.2. Drive Stabilization of the Soft Robot

Soft robots manufactured by the above methods are generally round, and therefore, there is a phenomenon of forward rotation during driving or backward rotation during the restoration stage of the soft jumping mechanism. To solve the problem and improve the performance of the soft jumping robot, the front part of the soft jumping robot is equipped with silicon structures in the back and middle, as shown in Figure 14. The soft jumping robot has 0.062 m long, 0.096 m wide, and 0.051 m high. The weight of the soft jumping robot is 0.131 kg.

The attached silicon structures can improve the phenomenon of the rolling forward because of inertia as shown in Figure 15, where the robot can land stably after jumping. The soft jumping robot's back foot does not pull the whole body due to friction with the floor surface of the silicon structure in the middle; instead, the rear foot is pulled to the main body side.

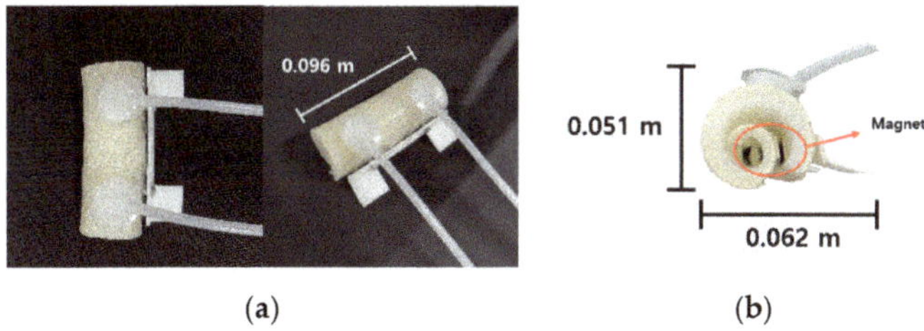

(**a**)   (**b**)

Figure 14. Final design of soft jumping robot. (**a**) Attachment position of front foot; (**b**) magnet for drive stabilization of soft jumping robot.

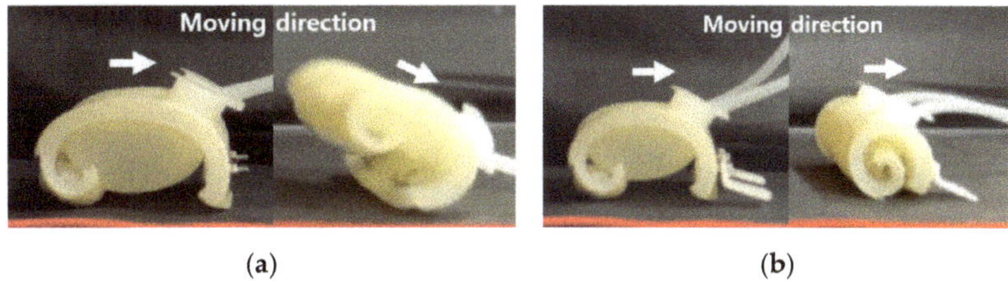

(**a**)   (**b**)

Figure 15. Comparing landing motion for mounting front foot; (**a**) Landing motion before mounting front foot. (**b**) landing motion after mounting front foot type.

7. Drive Test of Soft Robot

7.1. Performance Verification of the Soft Jumping Robots

To check the dynamic performance of the soft jumping robot, the various conditions of the robot were tested, such as magnet sizes, the amount of air, and the air injecting speed. The original jumping performance has experimented on the flat surface. In addition, the experiments at the various environmental conditions were also conducted to verify the usability of the robot.

The experiment was carried out on a rubber plate with the same specifications, as indicated in Table 3.

Table 3. Conditions for the robot used in the changing direction.

Magnet Size (m) (Length × Width × Thickness)	Amount of Air	Air Injecting Speed
0.01 × 0.02 × 0.003	0.04 L	0.0124 L/s

For the experiment, the soft jumping mechanism located in the lower half of the robot was operated separately. It is easy to switch between the left and right directions as shown in Figure 16.

The average rotation in the left direction excluding the highest and lowest values is $112.5° \pm 7.5°$ and $105.8° \pm 10.8°$ to the right. There was a small difference, but it was certainly able to change direction. Data on the experiments are provided in Figure 17.

The experiment on the jump performance of soft jumping robots was conducted by changing the magnet used to investigate the difference in performance, the amount of air injected, and the speed of air injected. The soft jumping robot used in the experiment used the same body.

An experiment was conducted to confirm whether the performance of the soft robot is improved according to the magnetic force installed in the driving part of the soft jumping robot. The maximum jump height and distance of the soft jumping robot were determined by driving the soft jumping robot 10 times from the same starting point. The conditions applied to the soft jumping robot are the same as those shown in Table 4.

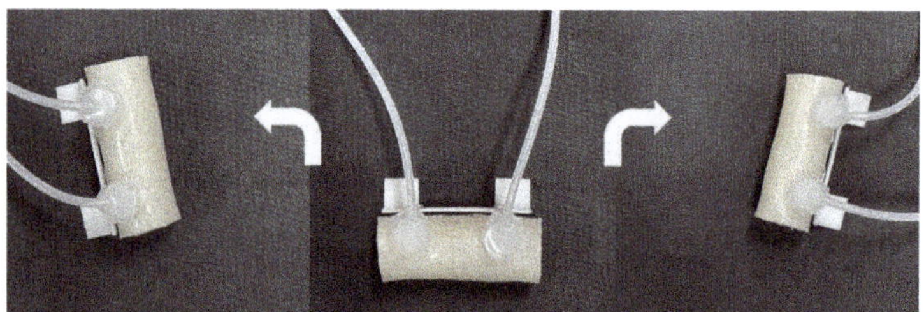

Figure 16. Rotation drive of the soft jumping robot.

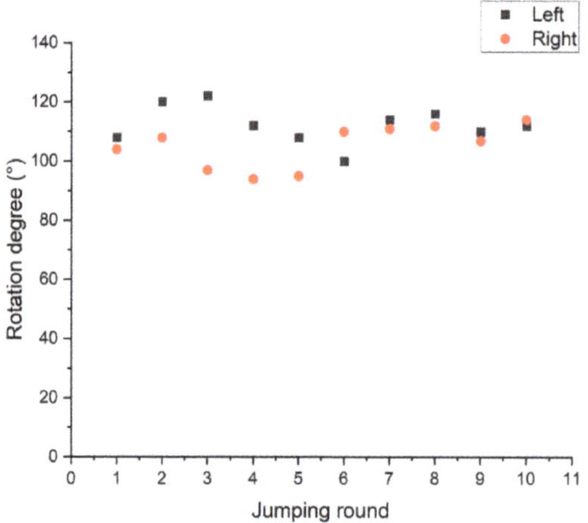

Figure 17. Result of changing direction test.

Table 4. Conditions for the robot used in the comparison of jump forces according to magnetic force test.

Magnet Size (m) (Length × Width × Thickness)	Amount of Air	Air Injecting Speed
0.01 × 0.02 × 0.003 0.01 × 0.015 × 0.003	0.04 L	0.0124 L/s

The result is shown in Figure 18. The performance of the jumping robot improved when the amount and speed of air injected were constant and the magnetic strength was high. In some cases, the experimental values were the same, so the graph was plotted as a single point.

The experiment was carried out while maintaining a constant injection speed to determine the effect of the amount of air supplied to the air chamber inside the soft jumping robot on the performance of the soft jumping robot. The amount of air injected is set at 0.035 L and 0.04 L, respectively, which does not add a burden on the drive. As in the previous experiment, the soft jumping robot was driven 10 times at the same starting point to determine the maximum jump height and distance of the soft jumping robot. The conditions applied to the soft jumping robot are summarized in Table 5.

The result is shown in Figure 19. The results show that the jumping robot's jump height and jump distance could be adjusted according to the amount of air injected. When the amount of injected air was 0.04 L, the robot jumped higher than 0.035 L, but 0.035 L jumped farther than 0.04 L. It is because after the yield point of the magnet (0.035 L) the remaining 0.005 L further expanded the air chamber to raise the robot's jumping angle.

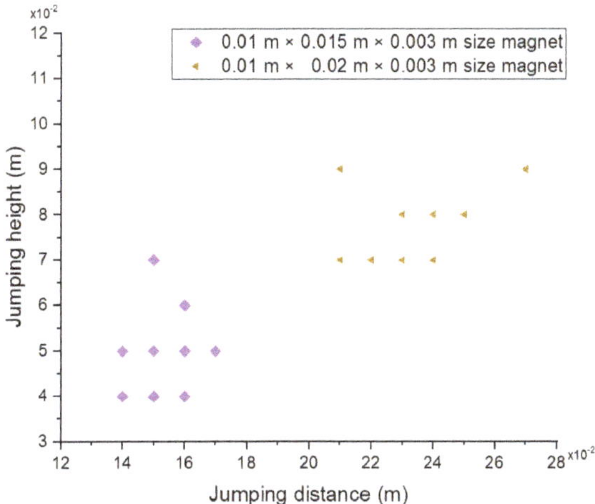

Figure 18. Result of jump forces depending on magnetic force test.

Table 5. Conditions for the robot used in the comparison of jump forces depending on the volume of air injected test.

Magnet Size (m) (Length × Width × Thickness)	Amount of Air	Air Injecting Speed
$0.01 \times 0.02 \times 0.003$	0.035 L 0.04 L	0.0124 L/s

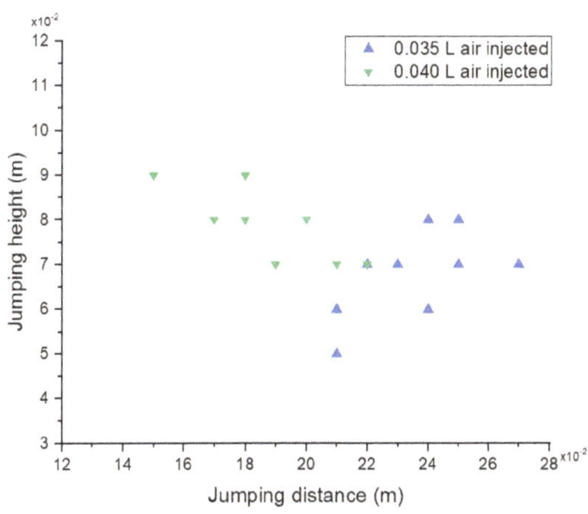

Figure 19. Result of comparison of the jump forces depending on the volume of air injected test.

The experiment was carried out by maintaining the magnet installed in the drive and the amount of air supplied to find out the effect of air injection speed on the performance of the soft jumping robot. Air injecting speed was set to 0.0124 L/s and 0.025 L/s, respectively, to the extent that it did not burden the driving part. As in the previous experiment, the maximum jump height and distance of the soft jumping robot were determined by driving the soft jumping robot 10 times. The conditions applied to the soft jumping robot are listed in Table 6.

Table 6. Conditions for the robot used in the comparison of jump forces depending on the air injecting speed.

Magnet Size (m) (Length × Width × Thickness)	Amount of Air	Air Injecting Speed
0.01 × 0.02 × 0.003	0.040 L	0.0124 L/s 0.025 L/s

The results are shown in Figure 20. The results indicate that the performance of the jumping robot can be improved when the amount of air injected, and the magnetic force of the permanent magnet installed are constant and the speed of air injected through the drive is high.

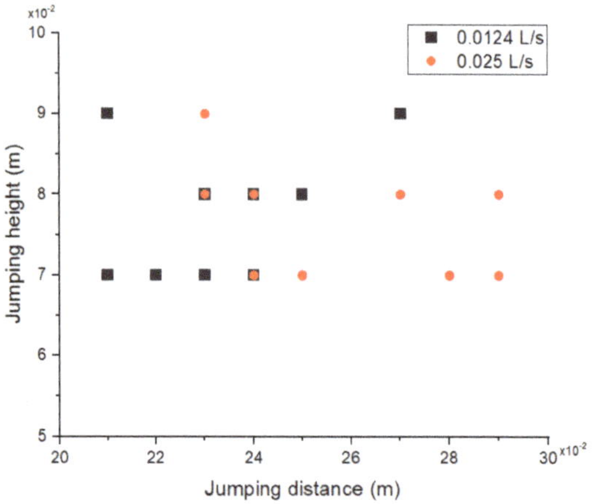

Figure 20. Comparison of jump forces depending on the air injecting speed.

The results of the driving tests show that the driving performance of soft-jumping robots using the soft-jumping mechanism produced in this paper are considerably affected by air injection. This is because the faster the air is injected from the magnetic yield point, the faster the energy is formed inside the air chamber, and the faster the driving occurs after the magnetic yield. Further, it was confirmed that the smaller the attached magnet, the lower is the driving performance of the soft jumping robot, and the more air it injects, the better is the performance.

7.2. Driving Tests under a Variety of Environments

The conditions of the soft robot used for this experiment are summarized in Table 3.

The experiment was conducted on a flat surface, a puddle, and 10° slope condition, as shown in Figure 21. The figure shows the jumping drive of a soft jumping robot on a flat surface. The jumping robot has a cycle of about 3 s, and when the soft jumping robot starts jumping the angle between the soft jumping robot and the ground is about 60°. It was possible to jump stably without being overturned.

Comparative driving experiments were conducted on pools and flatlands as shown in Figure 22 to confirm the jumping stability of the soft jumping robots. The driving performance of the soft jumping robot was lower than that of the flat ground because of the water resistance in the puddle; however, the deterioration of the driving performance did not affect the driving on the flat surface. It is confirmed that a soft jumping mechanism can be driven continuously for a long time without fail because it does not use complex structures or electronic components and does not interfere with moisture. In the puddle, the jumping height of the robot was about 0.07 m and the jumping distance was about 0.02 m when the driving condition was the same as Table 3.

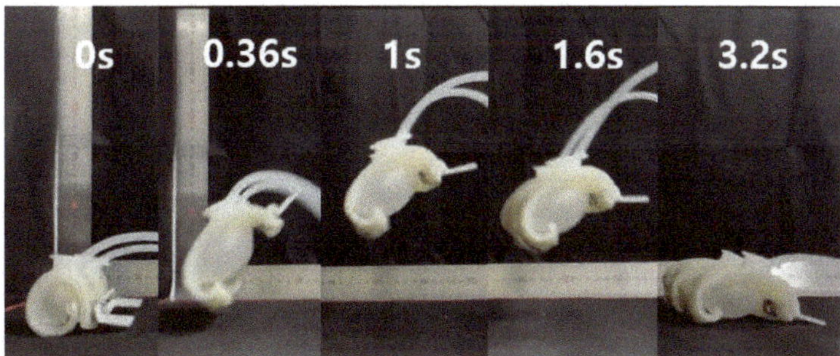

Figure 21. Drive motion of soft jumping robot.

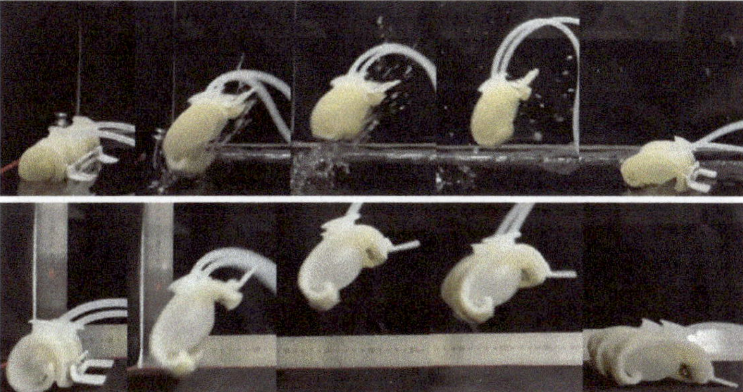

Figure 22. Comparison of the jumping motion of soft jumping robots on water pools and flat land.

Figure 23 shows the driving experiment conducted to evaluate the driving performance of a soft jumping robot on the 10° slope. However, it was found that the angle of the soft jumping robot's body at the beginning of jump on a flat surface was about 60°, while the 10° angle of the existing ground caused the soft jumping robot to drive at a higher angle than on a flat surface. Soft jumping robot jumped well on the 10° slope. When soft jumping robot driving in a puddle, lowering the jump force has been effective in increasing stability. In the 10° slope condition, the jumping height of the robot was about 0.07 m and the jumping distance was about 0.015 m when the driving condition was the same as Table 3. See Supplementary Materials.

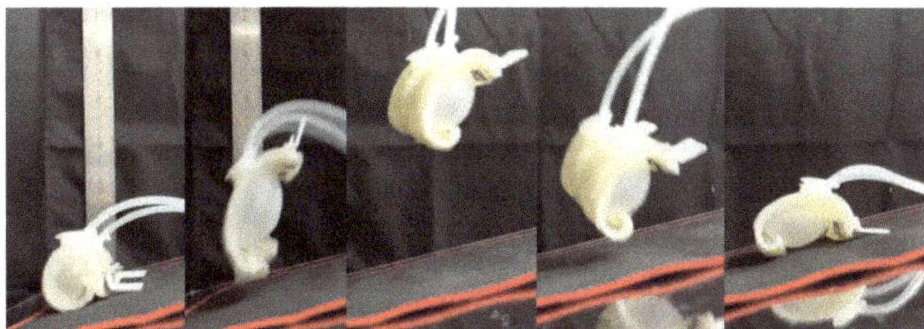

Figure 23. Jumping motion of soft jumping robots on 10° slope.

8. Conclusions

A soft morphing substrate manufacturing method was presented in this study. Using this method and magnetic yield point, we produced soft jumping robots that can store, release, and restore energy, and can change direction and jump in various environments to confirm the driving performance of soft jumping robots. Based on the results of the experiment, the thickness of the polymer structure was adjusted to suit the shape of the soft jumping robot, and the deformation rate of each part of the soft jumping robot was obtained. To confirm the driving performance of a soft robot with a soft jumping mechanism, 10 driving experiments were conducted. The experimental results indicate that the stronger the permanent magnet installed in the soft robot, the more air is injected, and the faster the air is injected, the higher is the performance under the same conditions. Soft jumping robots were tested on puddle and the $10°$ slope to investigate the possibility of driving under various environments. The results of the experiment confirmed that there was no overturning or damage when the robot moved in three different environments. Further, there was no difficulty in driving repeatedly.

In the future, adjusting the force of the magnet and the shape of the air chamber will reliably improve the jumping performance of the soft robot. At present, the pneumatic actuation part was not embedded in the soft robot. If the performance of the robot increases, the additional parts such as the pneumatic actuation part can be mounted into the soft jumping robot. It is necessary to add the additional parts compactly. Then the utilization of robots will increase even more.

Supplementary Materials: The following are available online at https://youtu.be/LJp1ld4btrE, Video: Soft Jumping Robot.

Author Contributions: G.-H.J.; investigation, G.-H.J. and Y.-J.P.; Conceptualization, methodology, validation, resources, writing—original draft preparation, writing—review and editing, project administration, funding acquisition. Both authors have read and agreed to the published version of the manuscript.

Funding: This research was supported by Basic Science Research Program through the National Research Foundation of Korea (NRF) funded by the Ministry of Education(2020R1I1A3073575) and by the National Research Foundation of Korea (NRF) Grant funded by the Korean Government (MSIT) (No.NRF-2016R1A5A1938472).

Institutional Review Board Statement: Not applicable.

Informed Consent Statement: Not applicable.

Data Availability Statement: Not applicable.

Acknowledgments: Kyu-Jin Cho and Jun-Young Lee; thank you for your advice on the bistable mechanism.

Conflicts of Interest: The authors declare no conflict of interest.

References

1. Bledt, G.; Powell, M.J.; Katz, B.; Carlo, J.D.; Wensing, P.M.; Kim, S. MIT Cheetah 3: Design and Control of a Robust, Dynamic Quadruped Robot. In Proceedings of the 2018 IEEE/RSJ International Conference on Intelligent Robots and Systems (IROS), Madrid, Spain, 1–5 October 2018; pp. 2245–2252.
2. Wooden, D.; Malchano, M.; Blankespoor, K.; Howardy, A.; Rizzi, A.A.; Raibert, M. Autonomous Navigation for BigDog. In Proceedings of the 2010 IEEE International Conference on Robotics and Automation, Anchorage, AK, USA, 3–8 May 2010; pp. 4736–4741.
3. Han, M.-W.; Kim, M.-S.; Ahn, S.-H. Shape Memory Textile Composites with Multi-Mode Actuations for Soft Morphing Skins. *Compos. Part B Eng.* **2020**, *198*, 108170. [CrossRef]
4. Umedachi, T.; Vikas, V.; Trimmer, B.A. Softworms: The Design and Control of Non-Pneumatic, 3D-Printed, Deformable Robots. *Bioinspir. Biomim.* **2016**, *11*, 025001. [CrossRef]
5. Zhang, Y.; Zhang, N.; Hingorani, H.; Ding, N.; Wang, D.; Yuan, C.; Zhang, B.; Gu, G.; Ge, Q. Fast-Response, Stiffness-Tunable Soft Actuator by Hybrid Multimaterial 3D Printing. *Adv. Funct. Mater.* **2019**, *29*, 1806698. [CrossRef]
6. Liu, Y.; Shaw, B.; Dickey, M.D.; Genzer, J. Sequential Self-Folding of Polymer Sheets. *Sci. Adv.* **2017**, *3*, e1602417. [CrossRef] [PubMed]

7.   Kim, H.-I.; Han, M.-W.; Song, S.-H.; Ahn, S.-H. Soft Morphing Hand Driven by SMA Tendon Wire. *Compos. Part B Eng.* **2016**, *105*, 138–148. [CrossRef]

8.   Shepherd, R.F.; Stokes, A.A.; Freake, J.; Barber, J.; Snyder, P.W.; Mazzeo, A.D.; Cademartiri, L.; Morin, S.A.; Whitesides, G.M. Using Explosions to Power a Soft Robot. *Angew. Chem. Int. Ed.* **2013**, *52*, 2892–2896. [CrossRef] [PubMed]

9.   Tolley, M.T.; Shepherd, R.F.; Karpelson, M.; Bartlett, N.W.; Galloway, K.C.; Wehner, M.; Nunes, R.; Whitesides, G.M.; Wood, R.J. An Untethered Jumping Soft Robot. In Proceedings of the 2014 IEEE/RSJ International Conference on Intelligent Robots and Systems, Chicago, IL, USA, 14–18 September 2014; pp. 561–566.

10.   Bartlett, N.W.; Tolley, M.T.; Overvelde, J.T.B.; Weaver, J.C.; Mosadegh, B.; Bertoldi, K.; Whitesides, G.M.; Wood, R.J. A 3D-Printed, Functionally Graded Soft Robot Powered by Combustion. *Science* **2015**, *349*, 161–165. [CrossRef] [PubMed]

11.   Wehner, M.; Truby, R.L.; Fitzgerald, D.J.; Mosadegh, B.; Whitesides, G.M.; Lewis, J.A.; Wood, R.J. An Integrated Design and Fabrication Strategy for Entirely Soft, Autonomous Robots. *Nature* **2016**, *536*, 451–455. [CrossRef] [PubMed]

12.   Yang, X.; Chang, L.; Pérez-Arancibia, N.O. An 88-Milligram Insect-Scale Autonomous Crawling Robot Driven by a Catalytic Artificial Muscle. *Sci. Robot.* **2020**, *5*, eaba0015. [CrossRef]

13.   Jung, G.; Casarez, C.S.; Jung, S.; Fearing, R.S.; Cho, K. An Integrated Jumping-Crawling Robot Using Height-Adjustable Jumping Module. In Proceedings of the 2016 IEEE International Conference on Robotics and Automation (ICRA), Stockholm, Sweden, 16–21 May 2016; pp. 4680–4685.

14.   Haldane, D.W.; Plecnik, M.; Yim, J.K.; Fearing, R.S. A Power Modulating Leg Mechanism for Monopedal Hopping. In Proceedings of the 2016 IEEE/RSJ International Conference on Intelligent Robots and Systems (IROS), Daejeon, Korea, 9–14 October 2016; pp. 4757–4764.

15.   Woodward, M.A.; Sitti, M. MultiMo-Bat: A Biologically Inspired Integrated Jumping–Gliding Robot. *Int. J. Robot. Res.* **2014**, *33*, 1511–1529. [CrossRef]

16.   Jiang, F.; Zhao, J.; Kota, A.K.; Xi, N.; Mutka, M.W.; Xiao, L. A Miniature Water Surface Jumping Robot. *IEEE Robot. Autom. Lett.* **2017**, *2*, 1272–1279. [CrossRef]

17.   Kenneally, G.; De, A.; Koditschek, D.E. Design Principles for a Family of Direct-Drive Legged Robots. *IEEE Robot. Autom. Lett.* **2016**, *1*, 900–907. [CrossRef]

18.   Noh, M.; Kim, S.; An, S.; Koh, J.; Cho, K. Flea-Inspired Catapult Mechanism for Miniature Jumping Robots. *IEEE Trans. Robot.* **2012**, *28*, 1007–1018. [CrossRef]

19.   Tang, Y.; Chi, Y.; Sun, J.; Huang, T.-H.; Maghsoudi, O.H.; Spence, A.; Zhao, J.; Su, H.; Yin, J. Leveraging Elastic Instabilities for Amplified Performance: Spine-Inspired High-Speed and High-Force Soft Robots. *Sci. Adv.* **2020**, *6*, eaaz6912. [CrossRef]

20.   Liu, G.-H.; Lin, H.-Y.; Lin, H.-Y.; Chen, S.-T.; Lin, P.-C. A Bio-Inspired Hopping Kangaroo Robot with an Active Tail. *J. Bionic. Eng.* **2014**, *11*, 541–555. [CrossRef]

21.   Zaitsev, V.; Gvirsman, O.; Ben Hanan, U.; Weiss, A.; Ayali, A.; Kosa, G. A Locust-Inspired Miniature Jumping Robot. *Bioinspir. Biomim.* **2015**, *10*, 066012. [CrossRef] [PubMed]

22.   Wang, M.; Zang, X.; Fan, J.; Zhao, J. Biological Jumping Mechanism Analysis and Modeling for Frog Robot. *J. Bionic Eng.* **2008**, *5*, 181–188. [CrossRef]

23.   Ache, J.M.; Matheson, T. Passive Joint Forces Are Tuned to Limb Use in Insects and Drive Movements without Motor Activity. *Curr. Biol.* **2013**, *23*, 1418–1426. [CrossRef]

24.   Baek, S.-M.; Yim, S.; Chae, S.-H.; Lee, D.-Y.; Cho, K.-J. Ladybird Beetle—Inspired Compliant Origami. *Sci. Robot.* **2020**, *5*, eaaz6262. [CrossRef]

25.   Chen, Z.; Majidi, C.; Srolovitz, D.J.; Haataja, M. Tunable Helical Ribbons. *Appl. Phys. Lett.* **2011**, *98*, 011906. [CrossRef]

26.   Han, M.-W.; Ahn, S.-H. Blooming Knit Flowers: Loop-Linked Soft Morphing Structures for Soft Robotics. *Adv. Mater.* **2017**, *29*, 1606580. [CrossRef] [PubMed]

27.   Franinović, K.; Franzke, L. Shape Changing Surfaces and Structures: Design Tools and Methods for Electroactive Polymers. In Proceedings of the 2019 CHI Conference on Human Factors in Computing Systems, Glasgow, Scotland, UK, 4–9 May 2019; pp. 1–12.

28.   Kim, S.; Koh, J.; Cho, M.; Cho, K. Towards a Bio-Mimetic Flytrap Robot Based on a Snap-through Mechanism. In Proceedings of the 2010 3rd IEEE RAS EMBS International Conference on Biomedical Robotics and Biomechatronics, Tokyo, Japan, 26–29 September 2010; pp. 534–539.

29.   Shahinpoor, M. Biomimetic Robotic Venus Flytrap (Dionaea Muscipula Ellis) Made with Ionic Polymer Metal Composites. *Bioinspir. Biomim.* **2011**, *6*, 046004. [CrossRef] [PubMed]

30.   Shepherd, R.F.; Ilievski, F.; Choi, W.; Morin, S.A.; Stokes, A.A.; Mazzeo, A.D.; Chen, X.; Wang, M.; Whitesides, G.M. Multigait Soft Robot. *Proc. Natl. Acad. Sci. USA* **2011**, *108*, 20400–20403. [CrossRef] [PubMed]

31.   Vokoun, D.; Beleggia, M.; Heller, L.; Šittner, P. Magnetostatic Interactions and Forces between Cylindrical Permanent Magnets. *J. Magn. Magn. Mater.* **2009**, *321*, 3758–3763. [CrossRef]

Article

Snake Robot with Driving Assistant Mechanism

Junseong Bae [1], Myeongjin Kim [1], Bongsub Song [1], Maolin Jin [2] and Dongwon Yun [1,*]

[1] Department of Robotics, Daegu Gyeongbuk Institute of Science and Technology (DGIST), Daegu 42988, Korea; bjs4578@dgist.ac.kr (J.B.); hambaf002@dgist.ac.kr (M.K.); doorebong@dgist.ac.kr (B.S.)

[2] Human-Centered Robotics Center, Korea Institute of Robotics & Technology Convergence, Pohang 791-941, Korea; mulimkim@kiro.re.kr

* Correspondence: mech@dgist.ac.kr; Tel.: +82-53-785-6219; Fax: +82-53-785-6209

Received: 28 September 2020; Accepted: 22 October 2020; Published: 24 October 2020

Abstract: Snake robots are composed of multiple links and joints and have a high degree of freedom. They can perform various motions and can overcome various terrains. Snake robots need additional driving algorithms and sensors that acquire terrain data in order to overcome rough terrains such as grasslands and slopes. In this study, we propose a driving assistant mechanism (DAM), which assists locomotion without additional driving algorithms and sensors. In this paper, we confirmed that the DAM prevents a roll down on a slope and increases the locomotion speed through dynamic simulation and experiments. It was possible to overcome grasslands and a 27 degrees slope without using additional driving controllers. In conclusion, we expect that a snake robot can conduct a wide range of missions well, such as exploring disaster sites and rough terrain, by using the proposed mechanism.

Keywords: snake robot; driving assistant mechanism; slope; dynamic analysis

---

1. Introduction

Biological snakes can overcome environmental irregularities such as rocky or grassy terrain, narrow spaces, and slopes by using different types of gait and body shapes [1,2]. Recently, research has been actively conducted to apply the biological locomotion of snakes through mimicry to robots [3–5]. Furthermore, many researchers studied new gait types for snake robots that have higher locomotion efficiency than a biological snake or novel locomotion properties [6,7]. The scale geometry of snake robots has been studied to stabilize their motion [8]. In addition, snake robots can be used to explore terrain or find missing people in disasters and can be utilized in various situations such as earthquake, fire site, and pipe inspections [9–11].

However, to overcome environmental irregularities, such as those that are found following disasters, snake robots need to use the optimal gait type according to the terrain, and a complex mathematical model to find the optimal gait type for each terrain is required [12–16]. In addition, a complicated semi-autonomous algorithm is needed to change the optimal gait type based on the terrain after receiving information about the terrain from additional sensors [17–19]. Furthermore, to support locomotion of snake robots in extreme environments, a mechanical design to overcome rough terrain, such as a wheel-based snake robot [20,21] or a skin drive snake robot [22], has to be considered. Especially in the case of a steep slope, a snake robot needs an additional sensor to receive information about the frictional force of the slope terrain, and the optimal gait type for the slope should be considered to prevent the snake robot from rolling down the steep slope, like Figure 1a. As a result, to overcome a steep slope, snake robots need to consider additional sensors and sophisticated algorithms to find the optimal gait type [23–25].

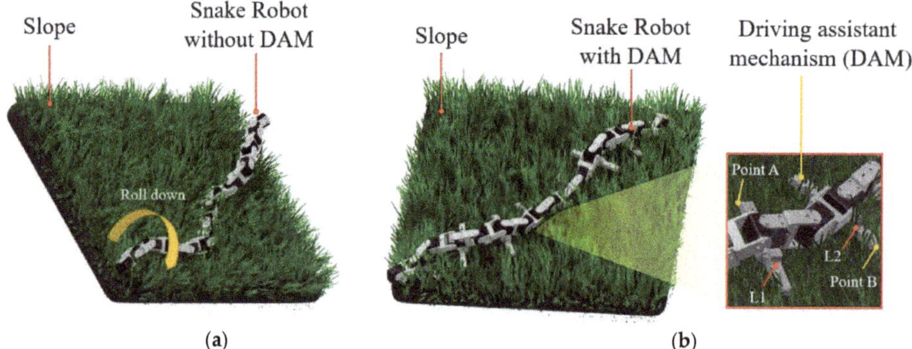

(**a**)   (**b**)

Figure 1. Conceptual diagram for locomotion of a snake robot (**a**) without a driving assistant mechanism (DAM) and (**b**) with a DAM.

In 2006, Shugen Ma et al. created a snake-like robot that was able to overcome a 20 degrees steep slope at a maximum forward speed of 0.2 m/s by applying the optimum winding angle to each joint [23]. In 2017, a research group demonstrated through simulation that a snake robot equipped with an encoder and inertia measurement unit (IMU) sensor could overcome slopes of from 5 to 25 degrees by predicting the angle of the slope through the sensor value. In this simulation, the gait type for the snake robot only considered rolling motion [24]. In addition, in 2020, DONGFANG LI et al. applied a motion planning algorithm and drive wheels to a multi-joint snake-like robot to overcome a slope of 7 degrees at a maximum speed of 0.4 m/s [25]. In conclusion, many researchers have been actively researching to find the optimal gait type for snake robots by using sensors and driving algorithms to overcome steep slopes.

There has been no research into the additional mechanisms or body shapes of a snake robot that can help snake robots to move efficiently on steep slopes, like those shown in Figure 1. In addition, the body shape of conventional snake robots is cylindrical or rectangular [1–6,9–15,17–22]. For this reason, when the snake robot moves sideways on the steep slope, the robot rolls down, and the stability is highly decreased. In conclusion, to operate a snake robot on a steep slope, it is necessary to study additional mechanisms to prevent the roll down motion of the snake robot.

In this paper, we propose a driving assistant mechanism (DAM) that assists snake robots with locomotion on steep slopes. A snake robot with a DAM is shown in Figure 1b. The DAM bears the weight of the snake robot. So, the DAM can help to prevent the roll down motion when the snake robot moves sideways on the steep slope. In addition, when a DAM is attached to a snake robot in the manner shown in Figure 1b, Link1 (L1) reaches to Point A, and Link2 (L2) reaches to Point B, distributing the weight of the snake robot to both sides. For this reason, the snake robot can overcome the real environment, such as bushes or gravel. As a result, the DAM can help to support the weight of the snake robot, which highly increases the driving stability. Furthermore, the driving algorithm for supporting the weight of the snake and the sensor for detecting surface information will be simplified due to the DAM, and the control efficiency will be highly increased compared to that of conventional snake robots due to the simplified algorithm and sensors. As a result, the DAM presented in this paper improves the snake robot's driving performance on steep slopes. For this reason, the DAM with a snake robot is expected to be used in various environments such as exploring terrain and saving lives in disaster situations. In some research about centipede-like robots, there are similar robot structures with a DAM, but there are significant differences. The DAM leg is an active actuator part that generates locomotion in centipede robots, while it is a passive structure that just prevents rolling in snake robots.

The contents of the paper are organized as follows. In Section 2, to apply the DAM to the snake robot, dynamic modeling of the DAM is conducted under a steep slope condition, and the design of the snake robot with the DAM is covered. In Section 3, we measure how many slope angles can be covered when we attach the DAM to a snake, and compare the locomotion performance through the locomotion experiments with and without a DAM in actual environments. Section 4 covers the conclusions and future work.

2. Materials and Methods

2.1. Design of the Snake Robot with a DAM

The DAM was connected to the snake robot's motor frame as shown in Figure 2a. As shown in Figure 2b, the DAM has a linear shape and is attached to the servo motor that rotates in the pitch axis.

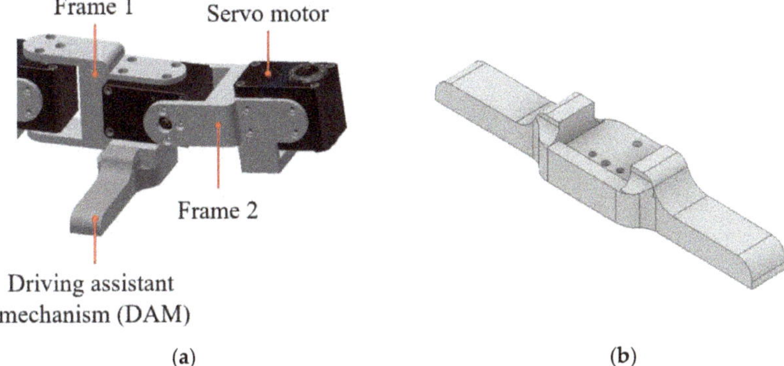

Figure 2. Modeling of a snake robot with a DAM. (**a**) Snake robot with a DAM. (**b**) Design of the DAM.

Before manufacturing the DAM, we performed a finite element analysis (FEA) model as shown in Figure 3 to confirm whether the DAM could withstand the gravity of the snake robot. The DAMs receive the gravity of the snake robots when the snake robot performs a vertical wave motion. We supposed that a force of 14 N is uniformly applied to the bottom of just one DAM. As shown in Figure 3a, the part receiving the high force is the center point of the DAM, and the maximum value is 2.025 MPa. The DAM is composed of high impact polystyrene (HIPS) material (Z-HIPS, Zortrax, Olsztyn, Poland), and the allowable capacity of this material is 29.3 MPa, so it is clear that the material is appropriate for making the DAMs. In Figure 3b, the portion with the maximum position is the end of the DAM, and the largest value is 0.04 mm. When the snake robot performs a vertical motion on the slope, the DAM is not heavily bent or damaged, so it can be confirmed that the stability of the DAM is enough to perform the vertical motion.

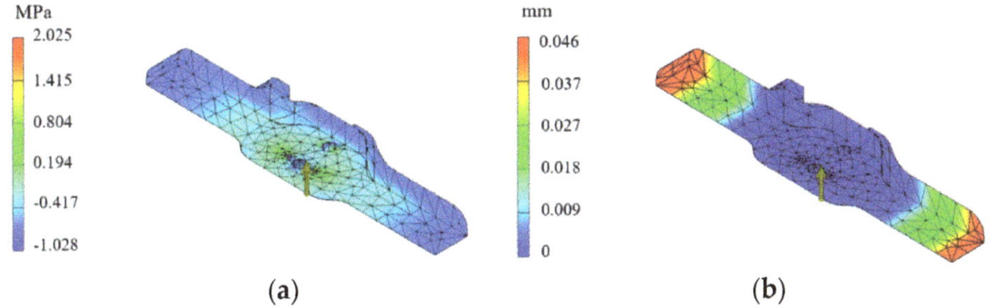

Figure 3. Analysis models of the DAM using a finite element analysis (FEA). (**a**) Stress analytical result. (**b**) Displacement analytical result.

2.2. Dynamic Modeling of Driving Assistant Mechanism

When a snake robot without a driving assistant mechanism performs locomotion on a slope, the reaction force of the force applied to the slope generates a torque on the snake robot, causing the snake robot to rotate and roll down. If a DAM is attached to the snake robot, the rolling down can be prevented and the stability of the snake robot can be increased. To test this, the snake robot made a vertical motion on a slope and the angle of the slope was gradually increased in order to confirm the maximum slope angle that snake robot could overcome.

As shown in Figure 4, the snake robot formed vertical waves on a slope, and the M1(servo motor 1) to M16 were connected by crossing each other at 90 degrees. In this case, the odd-numbered servo motors rotated around the yaw axis, and the even-numbered servo motors rotated around the pitch axis. When the snake robot performed the vertical motion, the odd-numbered servo motors were fixed at 0 degrees, and the snake robot formed a vertical wave by using even-numbered servo motors. Each rotation had a delay time of about 15 ms, and successive rotating occurred. After the rotation of M16, the same process was repeated to create a vertical motion.

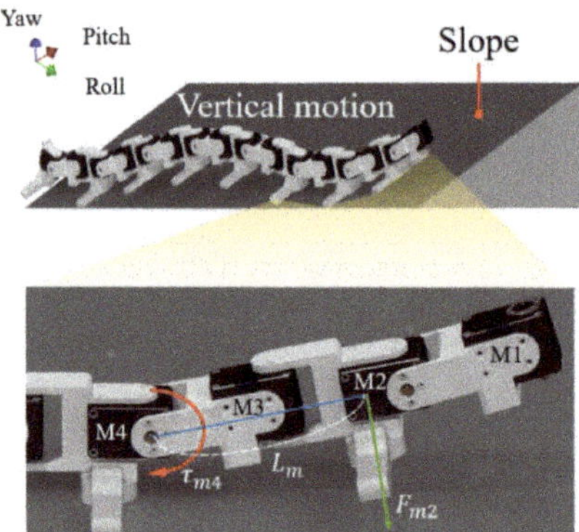

Figure 4. Configuration for dynamic modeling of the snake robot with a DAM.

Looking at M1–M4, the motors formed a vertical wave before the motors' operation. If M4 is operated clockwise, the torque is applied clockwise (as showed by the red arrow) and M1–M3 also turn clockwise. The DAM of M2 contacts the slope, then, an action force and reaction force are generated. At this time, the generated reaction force acts uniformly on the DAM, as shown in Figure 5a. This force can be substituted to a resultant force that acts on the center of the DAM and makes the torque counterclockwise in the roll axis. In contrast, the torque caused by gravity acting on the snake robot works clockwise in the roll axis, and the magnitude of the two torques determines whether the snake robot is rolled down or not.

As shown in Figure 5b, when it lifts from the slope, the snake robot has initial angular velocity and rotational kinetic energy that is counterclockwise and clockwise to the torque caused by gravity acting on the snake robot. The magnitude of the opposing energy and torque determines whether the snake robot is rolled down. When the rotational kinetic energy is consumed by the torque due to gravity, whether the snake robot rolls down is determined based on the position of the center of gravity of the snake robot. The snake robot turns clockwise due to the gravity torque when the center of gravity is located on the right side of the Z-axis indicated by the dotted line in Figure 5c, and the snake robot becomes an initial state. On the other hand, the snake robot turns counterclockwise due to the gravity torque when the center of gravity is located on the left side of the Z-axis, and the snake robot rolls down. The condition of the snake robot rolling down is shown in Equation (1). From Figure 5c, θ_2 can be derived with the geometry of the snake robot using Equation (2).

$$\Delta\theta > \frac{\pi}{2} - \theta_1 - \theta_2 \tag{1}$$

$$\theta_2 = \tan^{-1}\frac{L_{cm}}{L_2} \tag{2}$$

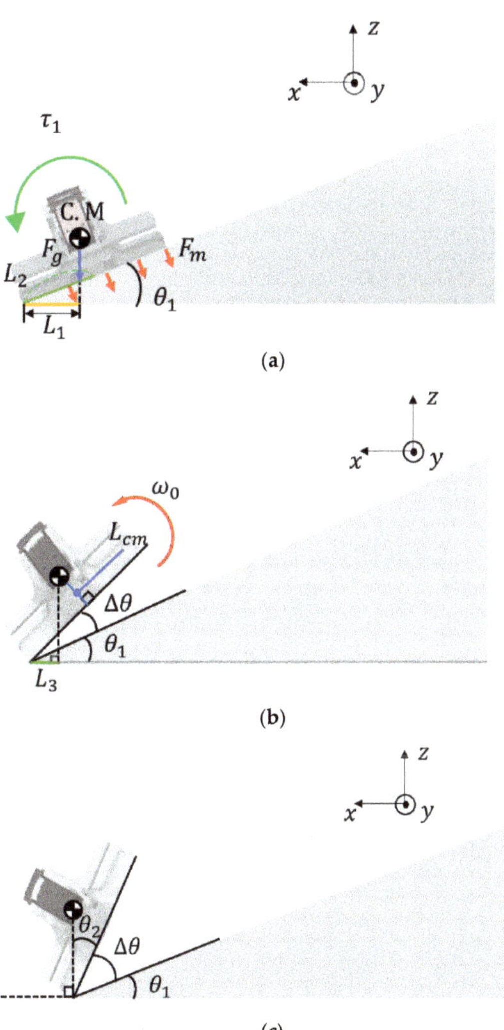

Figure 5. Free body diagram of the dynamic modeling. (**a**) The force which acted on DAM and slope. (**b**) Freebody diagram when the DAM lifted. (**c**) The moment when the center of mass located on the z-axis.

To derive the value of the maximum slope angle that the snake robot can overcome, a dynamic modeling was established by Equation (1) and the parameters in Table 1. In the case of vertical motion, the average number of DAM contacts with the ground during vertical motion was derived from experiments. As a result, total gravity of the snake robot could be dispersed in two DAMs. If τ_1 is defined as the sum of the torque due to gravity F_g and the torque due to the reaction force F_{m3} of F_m which is the acting force due to the rotation of M4, the Equations (3)–(5) can be developed as below.

$$F_g = \frac{16mg}{2} \tag{3}$$

$$F_{m3} = F_{m2} = LF_m \tag{4}$$

$$\tau_1 = F_{m3}L_2 - F_gL_1 \tag{5}$$

Table 1. Parameters for dynamic modeling.

Symbol	Description	Value	Unit
L	Length of the DAM (Length of the Servo Motor)	0.09, 0.12, 0.15, 0.18 (0.045)	m
L_{cm}	Y coordinate of the one module's center of mass	0.034	m
L_1	Distance between rotational axis and gravity force	-	m
L_2	Distance between rotational axis and action force	-	m
L_3	Distance between rotational axis and gravity force	-	m
θ_1	Degree of the inclined plane	-	rad
m	Mass of the one module of the snake robot	0.175	kg
I	Moment of inertia without DAM (with DAM)	0.00097 (0.0012)	kg·m^2
K_T	Motor torque constant	1.23	A/N·m
i_0	No load current	0.07	A
g	Acceleration of gravity	9.81	m/s^2
F_{m2}	Magnitude of substitution force of F_m	-	N
F_{m3}	Magnitude of reaction force of F_{m2}	-	N

When M2 receives τ_1 during Δt, angular velocity is generated, which can be written as Equation (6).

$$\tau_1 \Delta t = I\omega_0 \tag{6}$$

As described above, the snake robot has a rotational kinetic energy, $\frac{1}{2}I\omega_0^2$ caused by the initial angular velocity, and the torque, $\int_{\theta_1}^{\theta_1+\Delta\theta} F_g L_3 d\theta$ due to gravity acts to prevent the rolling the snake robot down until the center of mass of the snake robot reaches the Z-axis. Equation (7) is the condition in which the center of mass of the snake robot is located to the left side of the Z-axis. This condition is the same as that of the snake robot rolling down. Equation (8) is the result of substitution of the variables in the Equation (7). From Equation (8), the maximum slope angle can be theoretically obtained.

$$\frac{1}{2}I\omega_0^2 > \int_{\theta_1}^{\theta_1+\Delta\theta} F_g L_3 d\theta \tag{7}$$

$$\frac{1}{2}I\omega_0^2 > \frac{16mg}{2} \int_{\theta_1}^{\theta_1+(\frac{\pi}{2}-\theta_1-\theta_2)} \left(\frac{L}{2}\cos\theta - L_{cm}\sin\theta\right)d\theta \tag{8}$$

Using a gyro sensor attached to M2, the angle of the roll, pitch, and yaw axis with respect to time were measured, thereby obtaining the Δt value used in the dynamic modeling. In addition, the torque value used in dynamic modeling was obtained by measuring the current through experimentation in a similarly configured environment. The value of the torque was determined by the motor torque constant of the servo motor, the no-load current value, and the current at specific times as shown in Figure 6.

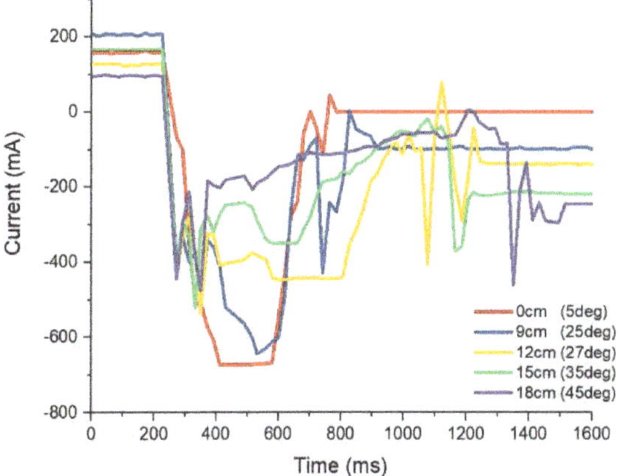

Figure 6. Graph of the current with various lengths of DAMs.

2.3. Experiment of Slope Condition

Figure 7 shows the experiment configuration for verifying dynamic modeling. M4 operated when the snake robot formed a vertical wave on a slope. In the state shown on the left in Figure 7, the M4 rotated by 30 degrees, and the DAM of the M2 pushed the bottom of the slope. Then, an acting force and a reaction force were generated. The same procedure was performed after raising the slope angle, and the maximum slope angle was measured. The maximum slope angles were measured when the DAM was not equipped and when the DAM was equipped with lengths of 9 cm, 12 cm, 15 cm, and 18 cm to confirm the influence of the length of the DAM.

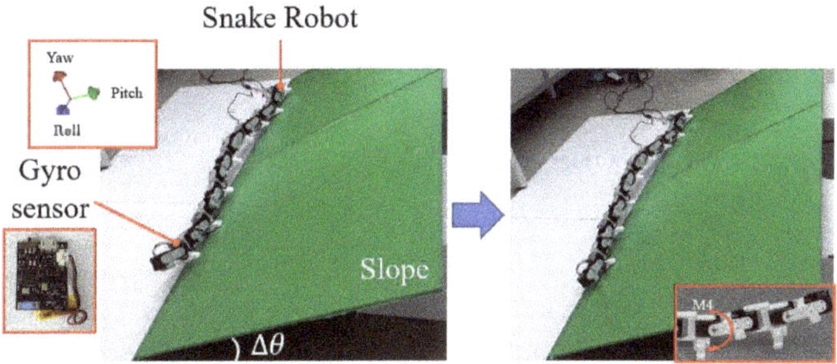

Figure 7. Experiment configuration for verifying dynamic modeling.

2.4. Locomotion Experiment for Various Gait Types

Snake robots should have various gait types to overcome rough terrain. In this section, we present the results of measurements on the locomotion performance when the snake robot conducted vertical motion, side winding motion, and sinus lifting motion on flatland. Each gait type of the snake robot is as shown in Figure 8.

The configuration for the experiment is as shown in Figure 9. The snake robot was controlled by a control PC, and a locomotion command was transmitted by a communication board (U2D2, Robotis). The distance from the start point to the end point was 1.5 m, and the time required for reaching the end point was measured, and the speed of each motion was measured. The locomotion speed when no DAM was attached and when the 12 cm DAM was attached were compared.

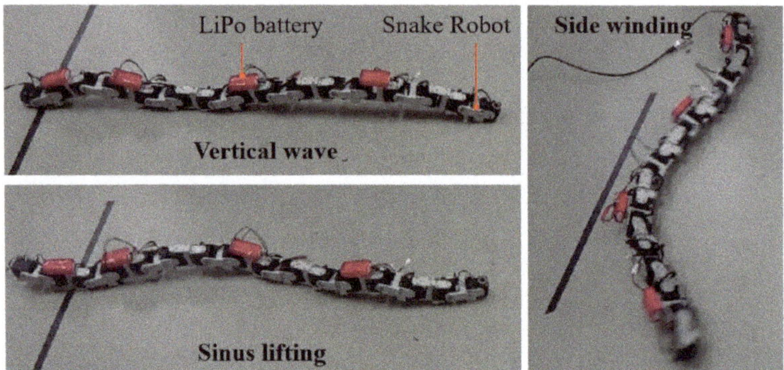

Figure 8. The snake robot's gait types on flatland.

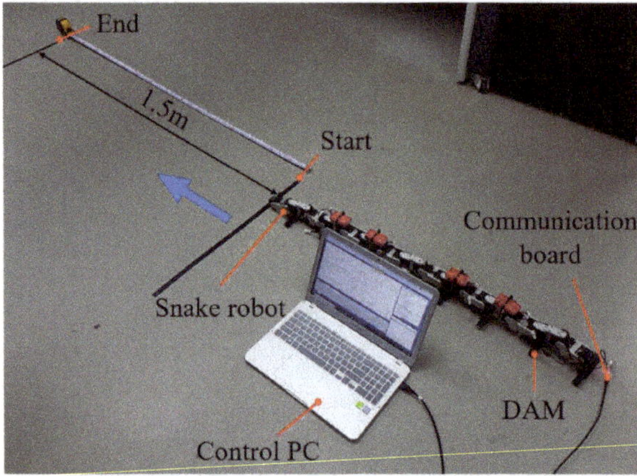

Figure 9. Configuration for flatland experiment.

2.5. Friction Experiment of the DAM and the Frame

The experiment measured the friction coefficient of the DAM and the frame (materials: Aluminum 6061) that connects the servo motors of the snake robot because these friction properties may affect the locomotion performance. A y-stage that can move the object uniformly and a 1 kg load cell were used in the experiment. The y-stage that was attached to the load cell pushed the DAM and frame with 200 g weight. The friction coefficients of the DAM and the frame were obtained by the fundamental friction equation for the friction force, normal force, and external force that were caused by the y-stage. The experiment's setup is shown in Figure 10.

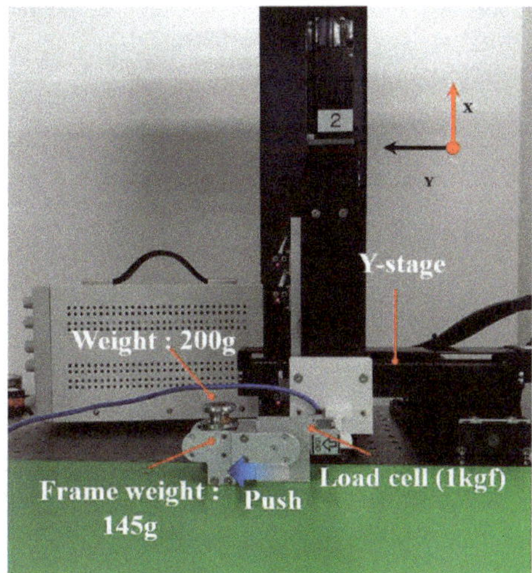

Figure 10. Experiment setup for measuring the friction coefficient.

2.6. Locomotion Experiment for Various Terrains

The environment used for testing locomotion performance on a slope is shown in Figure 11. The snake robot was placed on a 27 degree slope, and the slope was composed of grassland, which interrupted the locomotion of the snake robot. The purpose of this experiment was to confirm that the DAM works effectively to overcome the slope terrain without extra driving algorithms and sensors. For this purpose, we analyzed whether the snake robot rolls down when performing vertical motion, side winding motion, and sinus lifting motion on a slope.

Figure 11. Experiment environment in rough terrain.

An experiment on a 12 degrees slope that had obstacles was conducted for verifying the ascending performance of the DAM, as shown in Figure 12a. There were two types of obstacles used for testing the role of the DAM. The first obstacles were gravel, branches, and grass, the second obstacles were a small rock and large rock, as shown in Figure 12b,c. In this experiment, the vertical motion was used because vertical motion is suitable for ascending a slope.

(a)   (b)   (c)

Figure 12. Ascending environments composed of obstacles. (**a**) 12 Degree slope environments. (**b**) The experiments environments which have the grass, branches and gravel. (**c**) The experiments environments which have the small and huge rock.

An experiment on a 30 degrees slope that had various obstacles was performed for verifying the descending performance of the DAM, as shown in Figure 13. There were various obstacles such as gravel, a small rock, a large rock, and branches with soil. In this experiment, the sinus lifting motion was used because it is a stable motion. In addition, the experiment on the surface that is composed of gravel was conducted for verifying the robustness of the performance of the DAM, as shown in Figure 14. In this experiment, both the vertical motion and sinus lifting motion were used for testing the locomotion performance.

Figure 13. Descending environment composed of obstacles.

Figure 14. Gravelly field for testing the robustness of the performance of the DAM.

3. Results

3.1. Experiment Results of Slope Condition

A graph comparing the theoretical values with experiment values is shown in Figure 15. The maximum error between the theoretical values and the experiment values was 49% when the DAM length was 12 cm, and the minimum error value was 25% when the DAM length was 18 cm. As the length of the DAM increased, the maximum slope angle increased. The theoretical values and the experiment values were compared using a trend line and the method of least squares. As a result of the least squares method, the gradient of the theoretical values' trend line was 5.4, and the gradient of the experiment values' trend line was 6.8. It can be confirmed that the error of the ratio of the maximum slope angle was 20.6%.

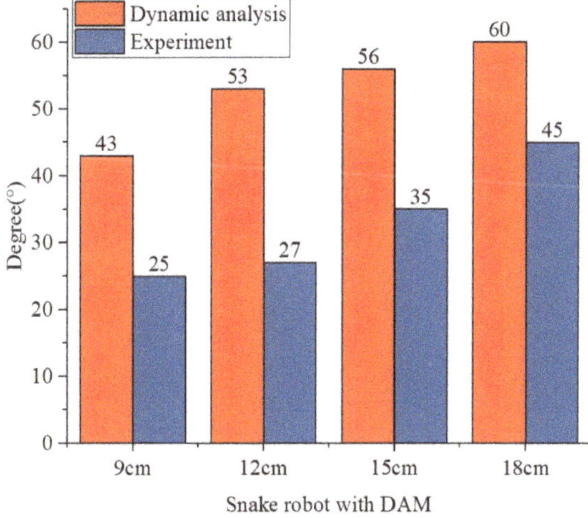

Figure 15. Dynamic modeling values and experiment values in the slope condition.

The value of the moment of inertia was measured using the I-property function of Autodesk inventor, and that may differ from the real value. The dynamic modeling did not consider the friction characteristics of the DAM and the surface. In addition, in the process of measuring the current or measuring Δt, the values of the sensor may have differed from the real values. These issues may have caused the error between the theoretical values and the experiment values. In the future, it is possible that the difference between the dynamic modeling and the experiment can be reduced by taking these factors into consideration.

3.2. Locomotion Experiment Results for Various Gait Types

The average value of three repetitions of the experiment results with and without the DAM are shown in Figure 16. When the DAM was not attached, the average value of the locomotion speed of the vertical motion was 4.44 cm/sec, the sinus lift motion was 4.14 cm/sec, and the sidewinding motion was 20.49 cm/sec. When a 12 cm DAM was attached to the snake robot, the average value of the locomotion speed of the vertical motion was measured at 6.34 cm/sec, the sinus lift motion was 8.01 cm/sec, and the side winding motion was 26.5 cm/s. As a result of attaching the DAM, the locomotion speed of the vertical motion increased by 29.96%, sinus lifting motion increased by 48.31%, and side winding motion increased by 22.6%. In this way, it was confirmed that the DAM plays a role not only in stable locomotion on a slope but also in increasing locomotion speed on flat terrain.

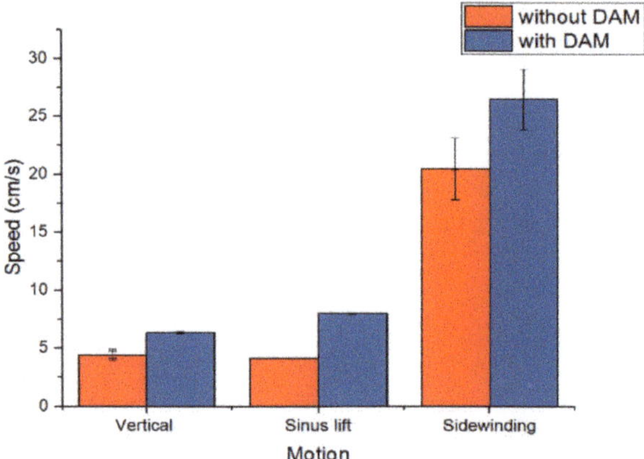

Figure 16. Locomotion speed for three movement types when the DAM was attached and when the DAM was not attached.

3.3. Friction Experiment of the DAM and the Frame

Each mean friction coefficients of the DAM and the frame were 0.050 and 0.044, respectively, as a result of the three repetitive experiments as shown in Figure 17. The difference between the two friction coefficients was 0.006, so we confirmed that the DAM did not change the friction property.

3.4. Locomotion Experiment Results on Various Terrains

When the DAM was not attached, the snake robot rolled down when performing the vertical motion, side winding motion, and sinus lifting motion due to unstable terrain. On the other hand, when the 12 cm DAM was attached, the snake robot conducted the same three locomotion types well in the same environment. The results are shown in Figures 18 and 19. In the case of vertical motion, the robot was not rolled and was able to go down the slope as shown in Figure 18. Next, we confirmed that both side winding motion and sinus lifting motion did not roll down on the slope and went down stably.

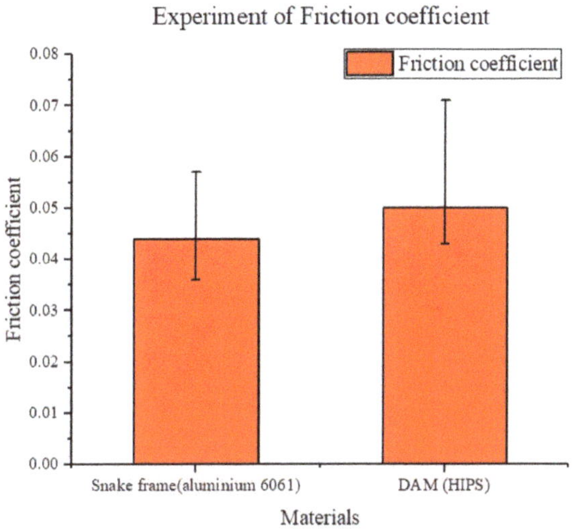

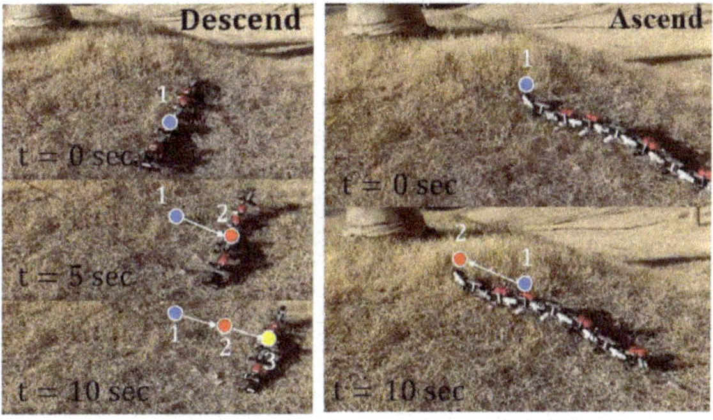

Figure 17. Friction coefficient result.

Figure 18. Performance of vertical motion with the DAM in rough terrain.

Figure 19. Performance of side winding and sinus lifting motion with the DAM in rough terrain.

The experiment on the 12 degrees slope that had obstacles was conducted for verifying the ascending performance of the DAM. In the case of the snake robot without the DAM, the snake robot rolled down when it hit the obstacles, as shown in Figure 20. In the case of the snake robot with the DAM, the snake robot ascended the 12 degrees slope that had branches and gravel without rolling down, as shown in Figure 21. In addition, the snake robot ascended the 12 degrees slope that had rocks by pushing the rocks with the DAM, as shown in Figure 22.

Figure 20. Performance of the ascending motion without the DAM.

Figure 21. Performance of the ascending motion with the DAM in the branches and gravel environment.

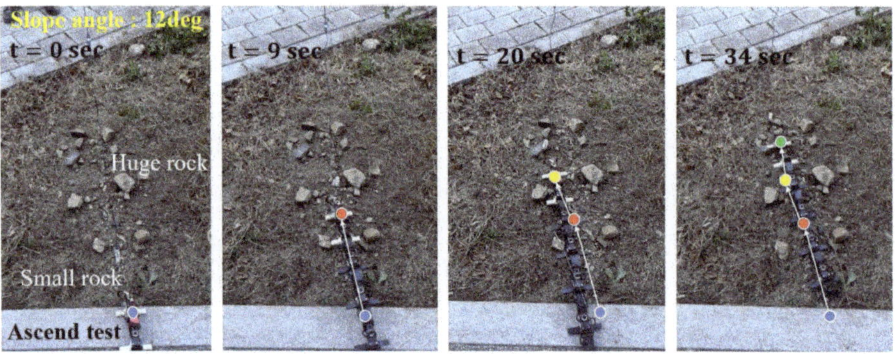

Figure 22. Performance of the ascending motion with the DAM on the rocks.

The experiment on the 30 degrees slope that had various obstacles was performed to verify the descending performance of the DAM. In the case of the snake robot without the DAM, the snake robot rolled down when it hit the obstacles, as shown in Figure 23. In the case of the snake robot with the DAM, the snake robot descended the 30 degrees slope that had branches, gravel, and rocks without rolling down, as shown in Figure 24.

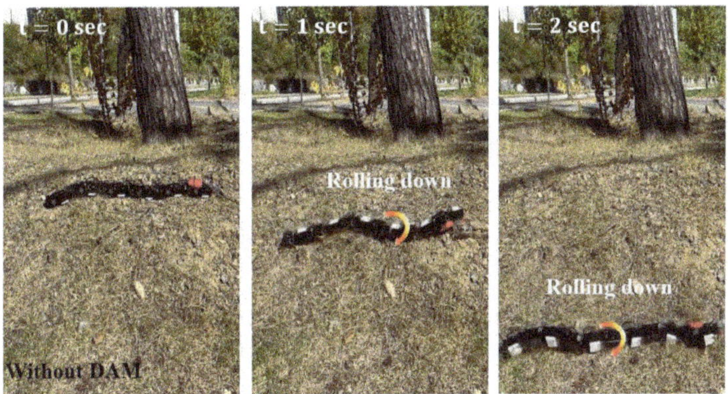

Figure 23. Performance of the descending motion without the DAM on the slope.

Figure 24. Performance of the descending motion with the DAM on the slope.

The experiment on the surface that was composed of gravel was conducted to verify the robust performance of the DAM. In the case of the snake robot without the DAM, the snake robot rolled down when it hit the gravel, as shown in Figure 25a,b, when the snake robot performed vertical motion and sinus lifting motion. With vertical motion of the snake robot with the DAM, the snake robot moved forward, but it got caught in the gravelly field, as shown in Figure 26a. In the sinus lifting motion of the snake robot with the DAM, the snake robot moved forward without catching in the gravelly field, as shown in Figure 26b, because the DAM pushed the gravel with the propulsion of the snake robot.

Figure 25. Performance of the robustness without the DAM on the gravelly field. (**a**) The vertical motion with DAM in the gravelly field. (**b**) The sinus lifting motion with DAM in the gravelly field.

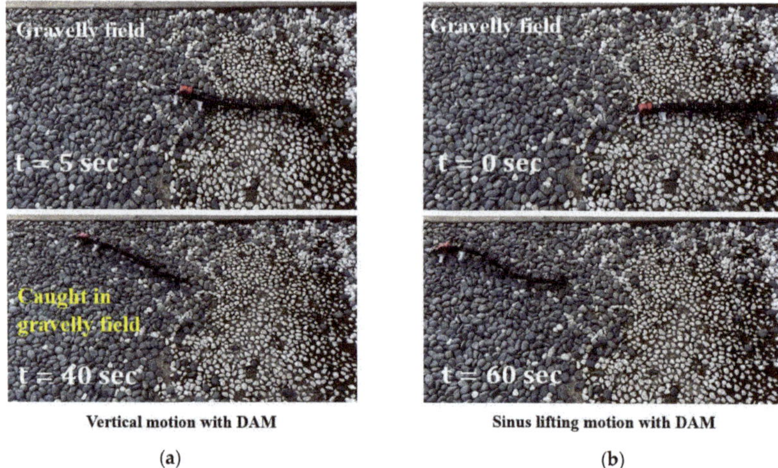

Figure 26. Performance of the robustness with the DAM on the gravelly field. (**a**) The vertical motion with DAM in the gravelly field. (**b**) The sinus lifting motion with DAM in the gravelly field.

4. Discussion and Conclusions

In this paper, we showed that when the driving assistant mechanism (DAM) is attached to a snake robot, it can perform locomotion on rough terrain such as a slope or grasslands without additional driving algorithms and sensors. In the slope environment, vertical motion and side winding motion can be used in descending the slope and vertical motion can be used in ascending the slope. Sinus lifting motion can be used in transverse motion of a slope environment. However, it is impossible for the DAM alone to overcome all terrain. In other words, if the DAM, driving algorithm, and sensors are in harmony, higher slopes and rougher terrains can be overcome. In addition, the DAM can be studied in the future for various materials such as soft materials and shapes.

A dynamic analysis of a snake robot with a DAM for vertical motion was conducted in this paper. The results were compared with the experiment's results in a slope condition. The maximum error between the theoretical values and the experiment values was 49% when the DAM length was 12 cm, and the minimum error value was 25% when the DAM length was 18 cm. In addition, it was confirmed that as the length of the DAM became longer, the maximum slope angle that the snake robot could overcome became larger from both the dynamic analysis and the experiment. When the length of DAM was 18 cm, it was confirmed that the snake robot did not roll down on a 42 degrees slope. However, if the length of the DAM becomes longer, the advantage of the snake robot being able to move through narrow terrain is reduced, and there is a possibility that the operation of the snake robot's motor may be interrupted. Therefore, DAMs must be able to fold and unfold if they are going to be used in the field. In addition, it is expected that advanced dynamic modeling of the side winding motion and sinus lifting motion will help overcome rough terrain effectively.

In locomotion experiments, it was confirmed that the locomotion speed was increased by the DAM. The locomotion speed increased by 29.96%, 48.31%, and 22.6% in the vertical motion, sinus lifting motion, and side winding motion, respectively. We measured the friction coefficient to find out the reason for the improvement in the speed. However, there was not a distinct difference between DAM's friction coefficient and the frame's friction coefficient; the friction coefficients of the DAM and the frame were 0.050 and 0.044, respectively. The improvement in the speed may be caused by other changes of dynamic properties and the clear reason of the improvement can be researched in future work. Through experiments on rough terrain, such as on a slope and in a gravelly field, it was confirmed that the DAM improved the locomotion performance. The DAM prevented rolling of the snake robot down on the slope and generated propulsion by pushing on the obstacles with the DAM. In the slope environment, the vertical motion with the DAM is suitable for ascending slopes and the sinus lifting motion with the DAM is suitable for descending slope. One drawback that was found is that the DAM caught in obstacles, however, this can be overcome by the sinus lifting motion. In this experiment, we confirmed that the sinus lifting motion with the DAM has robustness in an environment that has obstacles.

Author Contributions: Conceptualization, J.B., M.K., B.S. and D.Y.; methodology, J.B. and M.K.; formal analysis, J.B.; investigation, J.B., M.J.; data curation, M.K.; writing—original draft preparation, J.B.; writing—review and editing, J.B., M.K., M.J., and D.Y.; visualization, J.B.; supervision, D.Y. and M.J.; project administration, D.Y.; funding acquisition, D.Y. and M.J.; All authors have read and agreed to the published version of the manuscript.

Funding: This material is based upon work supported by the Ministry of Trade, Industry and Energy (MOTIE, Korea) under Industrial Technology Innovation Program. No. 20003739.

Conflicts of Interest: The authors declare no conflict of interest.

References

1. Sanfilippo, F.; Azpiazu, J.; Marafioti, G.; Transeth, A.A.; Stavdahl, Ø.; Liljebäck, P. Perception-driven obstacle-aided locomotion for snake robots: The state of the art, challenges and possibilities. *Appl. Sci.* **2017**, *7*, 336. [CrossRef]

2. Yun, D. Development of a mobile robot mimicking the frilled lizard. *J. Mech. Sci. Technol.* **2018**, *32*, 1787–1792. [CrossRef]

3. Ariizumi, R.; Matsuno, F. Dynamic analysis of three snake robot gaits. *IEEE Trans. Robot.* **2017**, *33*, 1075–1087. [CrossRef]

4. Liljeback, P.; Pettersen, K.Y.; Stavdahl, Ø.; Gravdahl, J.T. Controllability and stability analysis of planar snake robot locomotion. *IEEE Trans. Autom. Control* **2010**, *56*, 1365–1380. [CrossRef]

5. Takemori, T.; Tanaka, M.; Matsuno, F. Gait design for a snake robot by connecting curve segments and experimental demonstration. *IEEE Trans. Robot.* **2018**, *34*, 1384–1391. [CrossRef]

6. Wang, X.; Zhang, Q.; Shen, D.; Chen, J. A Novel Rescue Robot: Hybrid Soft and Rigid Structures for Narrow Space Searching. In Proceedings of the 2019 IEEE International Conference on Robotics and Biomimetics (ROBIO), Dali, Yunnan, China, 6–8 December 2019; pp. 2207–2213.

7. Manzoor, S.; Khan, U.; Ullah, I. Serpentine and Rectilinear Motion Generation in Snake Robot Using Central Pattern Generator with Gait Transition. *Iran. J. Sci. Technol. Trans. Electr. Eng.* **2019**, *44*, 1–11. [CrossRef]

8. Huq, N.M.L.; Khan, M.R.; Shafie, A.A.; Billah, M.M.; Ahmmad, S.M. Motion Investigation of a Snake Robot with Different Scale Geometry and Coefficient of Friction. *Robotics* **2018**, *7*, 18. [CrossRef]

9. Liljebäck, P.; Pettersen, K.Y.; Stavdahl, Ø.; Gravdahl, J.T. A review on modelling, implementation, and control of snake robots. *Robot. Auton. Syst.* **2012**, *60*, 29–40. [CrossRef]

10. Selvarajan, A.; Kumar, A.; Sethu, D.; bin Ramlan, M.A. Design and Development of a Snake-Robot for Pipeline Inspection. In Proceedings of the 2019 IEEE Student Conference on Research and Development (SCOReD), Bandar Seri Iskandar, Malaysia, 15–17 October 2019; pp. 237–242.

11. Whitman, J.; Zevallos, N.; Travers, M.; Choset, H. Snake robot urban search after the 2017 Mexico City earthquake. In Proceedings of the 2018 IEEE international symposium on safety, security, and rescue robotics (SSRR), Philadelphia, PA, USA, 6–8 August 2018; pp. 1–6.

12. Chang, A.H.; Vela, P.A. Shape-centric modeling for control of traveling wave rectilinear locomotion on snake-like robots. *Robot. Auton. Syst.* **2020**, *124*, 103406. [CrossRef]

13. Manzoor, S.; Cho, Y.G.; Choi, Y. Neural oscillator based CPG for various rhythmic motions of modular snake robot with active joints. *J. Intell. Robot. Syst.* **2019**, *94*, 641–654. [CrossRef]

14. Yun, D.; Fearing, R.S. Cockroach Milli-Robot With Improved Load Capacity. *J. Mech. Robot.* **2019**, *11*, 1. [CrossRef]

15. Yun, D.; Kim, K.-S.; Kim, S.; Kyung, J.; Lee, S. Note: Dynamic analysis of a robotic fish motion with a caudal fin with vertical phase differences. *Rev. Sci. Instrum.* **2013**, *84*, 036108. [CrossRef] [PubMed]

16. Gong, C.; Tesch, M.; Rollinson, D.; Choset, H. Snakes on an inclined plane: Learning an adaptive sidewinding motion for changing slopes. In Proceedings of the 2014 IEEE/RSJ International Conference on Intelligent Robots and Systems, Chicago, IL, USA, 14–18 September 2014; pp. 1114–1119.

17. Koopaee, M.J.; Bal, S.; Pretty, C.; Chen, X. Design and Development of a Wheel-less Snake Robot with Active Stiffness Control for Adaptive Pedal Wave Locomotion. *J. Bionic Eng.* **2019**, *16*, 593–607. [CrossRef]

18. Sanfilippo, F.; Helgerud, E.; Stadheim, P.A.; Aronsen, S.L. Serpens, a low-cost snake robot with series elastic torque-controlled actuators and a screw-less assembly mechanism. In Proceedings of the 2019 5th International Conference on Control, Automation and Robotics (ICCAR), Beijing, China, 19–22 April 2019; pp. 133–139.

19. Transeth, A.A.; Leine, R.I.; Glocker, C.; Pettersen, K.Y.; Liljebäck, P. Snake robot obstacle-aided locomotion: Modeling, simulations, and experiments. *IEEE Trans. Robot.* **2008**, *24*, 88–104. [CrossRef]

20. Fjerdingen, S.A.; Liljebäck, P.; Transeth, A.A. A snake-like robot for internal inspection of complex pipe structures (PIKo). In Proceedings of the 2009 IEEE/RSJ International Conference on Intelligent Robots and Systems, St. Louis, MO, USA, 10–15 October 2009; pp. 5665–5671.

21. Mori, M.; Hirose, S. Three-dimensional serpentine motion and lateral rolling by active cord mechanism ACM-R3. In Proceedings of the IEEE/RSJ International Conference on Intelligent Robots and Systems, Lausanne, Switzerland, 30 September–4 October 2002; pp. 829–834.

22. McKenna, J.C.; Anhalt, D.J.; Bronson, F.M.; Brown, H.B.; Schwerin, M.; Shammas, E.; Choset, H. Toroidal skin drive for snake robot locomotion. In Proceedings of the 2008 IEEE International Conference on Robotics and Automation, Pasadena, CA, USA, 19–23 May 2008; pp. 1150–1155.

23. Ma, S.; Tadokoro, N. Analysis of creeping locomotion of a snake-like robot on a slope. *Auton. Robot.* **2006**, *20*, 15–23. [CrossRef]

24. Bing, Z.; Cheng, L.; Knoll, A.; Zhong, A.; Huang, K.; Zhang, F. Slope angle estimation based on multi-sensor fusion for a snake-like robot. In Proceedings of the 2017 20th International Conference on Information Fusion (Fusion), Xi'an, China, 10–13 July 2017; pp. 1–6.

25. Li, D.; Wang, C.; Deng, H.; Wei, Y. Motion Planning Algorithm of a Multi-Joint Snake-Like Robot Based on Improved Serpenoid Curve. *IEEE Access* **2020**, *8*, 8346–8360. [CrossRef]

Publisher's Note: MDPI stays neutral with regard to jurisdictional claims in published maps and institutional affiliations.

applied sciences

Article

A Miniature Flapping Mechanism Using an Origami-Based Spherical Six-Bar Pattern

Seung-Yong Bae [1], Je-Sung Koh [2] and Gwang-Pil Jung [1,*]

[1] Department of Mechanical and Automotive Engineering, Seoul National University of Science and Technology, Seoul 01811, Korea; qo1113@seoultech.ac.kr
[2] Department of Mechanical Engineering, Ajou University, Suwon-si 16499, Korea; jskoh@ajou.ac.kr
* Correspondence: gpjung@seoultech.ac.kr

Abstract: In this paper, we suggest a novel transmission for the DC motor-based flapping-wing micro aerial vehicles (FWMAVs). Most DC motor-based FWMAVs employ linkage structures, such as a crank-rocker or a crank-slider, which are designed to transmit the motor's rotating motion to the wing's flapping motion. These transmitting linkages have shown successful performance; however, they entail the possibility of mechanical wear originating from the friction between relative moving components and require an onerous assembly process owing to several tiny components. To reduce the assembly process and wear problems, we present a geometrically constrained and origami-based spherical six-bar linkage. The origami-based fabrication method reduces the number of the relative moving components by replacing rigid links and pin joints with facets and folding joints, which shortens the assembly process and reduces friction between components. The constrained spherical six-bar linkage enables us to change the motor's rotating motion to the linear reciprocating motion. Due to the property that every axis passes through a single central point, the motor's rotating motion is filtered at the spherical linkage and does not transfer to the flapping wing. Only linear motion, therefore, is passed to the flapping wing. To show the feasibility of the idea, a prototype is fabricated and analyzed by measuring the flapping angle, the wing rotation angle and the thrust.

Keywords: bio-inspired robot; micro aerial vehicle; flapping mechanism

Citation: Bae, S.-Y.; Koh, J.-S.; Jung, G.-P. A Miniature Flapping Mechanism Using an Origami-Based Spherical Six-Bar Pattern. *Appl. Sci.* **2021**, *11*, 1515. https://doi.org/10.3390/app11041515

Academic Editor: Nicola Pio Belfiore
Received: 22 December 2020
Accepted: 3 February 2021
Published: 8 February 2021

Publisher's Note: MDPI stays neutral with regard to jurisdictional claims in published maps and institutional affiliations.

1. Introduction

Flapping-wing micro air vehicles (FWMAVs) have been widely studied due to their agility and maneuverability in the air. These properties can be applied to the areas of reconnaissance for military and exploration missions for disaster areas. For example, they can easily approach tiny cracks at disaster areas or enemy camps without being recognized.

To enable successful working of FWMAVs, researchers have employed a variety of actuators such as DC motors [1–19], piezoelectric actuators [20–23], electromagnetic actuators [24–28], and electrostatic actuators [6,29].

Among these actuators, in this paper, we suggest a novel transmission mechanism to reduce the assembly process and wear problems for the DC motor-based FWMAVs. In the case of DC motor-based FWMAVs, most of them use linkage structures to transmit the motor's rotating motion to the wing's flapping motion. These transmitting linkages have involve the possibility of mechanical wear originating from the friction between relative moving components and requires an onerous assembly process owing to several tiny components. Types of the transmissions used for the DC motor-based FWMAVs are given in Table 1.

Previously, a variety of transmission mechanisms have been employed. Galinski et al. used a spatial mechanism based on the double Scotch yoke [15]. They proposed to use spherical type of double Scotch yoke to overcome drawbacks of the planar Scotch yoke. Oppenheimer et al. and Sahai et al. employed four-bar linkages to change the motor's rotating motion to the flapping motion [5,12]. Karasek et al. and Wagter et al. also made

41

flapping mechanisms using crank-rocker four-bar linkages [2,4]. Takahashi et al. and Lau et al. made flapping mechanisms using crank-sliders [1,9]. The sliders enabled the mechanism to have reciprocating motion of the wings. Ristroph et al. and Gaissert et al. applied crank-shaft mechanisms [8,11]. The crank-shaft mechanisms converted motors' rotating movement into an oscillating movement of the wings. Roshanbin et al. proposed a flapping robot, COLIBRI, by combining four-bar linkage and a slider-crank mechanism [16].

Table 1. Transmission mechanisms used for DC motor-based flapping-wing micro air vehicles (FWMAVs).

Mechanism Types	Possibility of Mechanical Wear	Difficulty of Assembly	Transmission Efficiency
Scotch-yoke [15]	Yes	Dexterity is required	High (rotating in one direction)
Crank-rocker [2,4,5,12]	Yes	Dexterity is required	High (rotating in one direction)
Crank-slider [1,8,9,11,16]	Yes	Dexterity is required	High (rotating in one direction)
String-based [3,7,13,17–19]	No	Dexterity is required	High (rotating in one direction)
Proposed mechanism	No	Easy	High (rotating in one direction)
Direct drive [10,14]	No	Easy	Low (vibrate back and forth)

To reduce the mass and complexity of the transmission structures, some researchers suggested a string-based flapping mechanism or direct-driven mechanisms [7]. Keennon et al. and Gong et al. used string-based flapping mechanisms [3,13]. KUbeetle [17–19] mixed four-bar linkages and a string-based pulley mechanism. The string-based transmission enabled the prevention of mechanical wear and fracture of moving parts. The direct-driven mechanisms enabled the extreme reduction in the mechanical structures for transmission [10,14]. The direct-driven mechanisms achieved a high lift-to-weight ratio by resonating the system.

In this paper, we suggest a novel transmission mechanism to reduce the assembly process and wear problems for the DC motor-based FWMAVs. To make this possible, two design approaches have been utilized: origami-based fabrication and geometrically constrained spherical six-bar linkage. The origami-based fabrication method reduces the number of relative moving components by replacing rigid links and pin joints with facets and folding joints, which shortens the assembly process and reduces wear problems [30,31]. The geometrically constrained spherical six-bar linkage enables the change in the motor's rotating motion to the linear reciprocating motion. Due to the property that every axis passes through a single central point, the constrained spherical six-bar linkage filters the rotating motion of the DC motor and passes only linear motion to the flapping wings. Using the proposed design approaches, a novel flapping mechanism is made as shown in Figure 1.

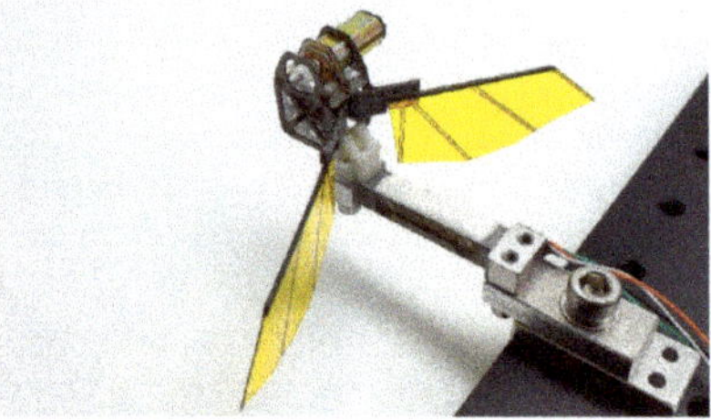

Figure 1. The developed prototype FWMAV based on flapping mechanism using a constrained spherical six-bar linkage.

This paper is organized as follows: The design section describes the concept of the flapping mechanism and the working process of the origami-based constrained spherical six-bar. To precisely investigate the mechanism and to select the proper geometric parameters, a three-dimensional kinematic model has been made. Furthermore, an experiment has been conducted to check the feasibility of the developed flapping mechanism by measuring the flapping frequency, angle, wing rotation angle and thrust.

2. Design

The overall design is shown in Figure 1. A DC motor is located at the top of the mechanism and the shaft is connected to the spur gear. The spur gear plays a role as a crank. The connecting rod connects the crank and the spherical six-bar linkage. The spherical six-bar linkage transmits the motor's motion to the flapping wing.

2.1. Flapping Mechanism

One important point is that only linear motion should be transmitted to generate the flapping motion of the wings. To make this possible, one end of the spherical six-bar linkage is fixed at the body frame as shown in Figure 2g. This constraint filters the rotating motion of the motor and transmits the linearly reciprocating motion only, which is shown in Video S1. By using this working principle, a successful flapping motion can be achieved without mechanical parts that cause friction problems.

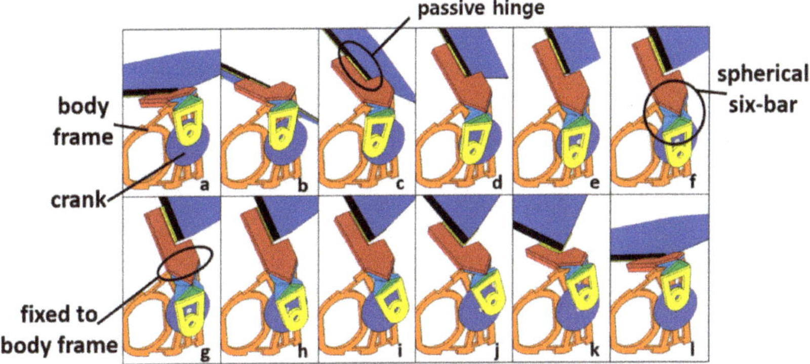

Figure 2. Snapshots of the moving process of the transmission. From (**a–f**), the mechanism shows upstroke. From (**g–l**), downstroke is shown.

To generate the thrust in the end, a wing rotating motion is necessarily required. To achieve the wing rotating motion, passive hinges are employed as shown in Figure 2c. The passive joints are positioned between the output of the transmission and the wings. The passive hinges are parallel to the spanwise direction and have a joint-stop to prevent excessive rotation. Overall, the system has three degrees of freedom—flapping actuation and two passive hinges for wing rotation.

Flapping motion is achieved by the constrained spherical six-bar linkage shown in Figure 3, which is the key enabling technology of the proposed mechanism. Generally, the spherical six-bar linkage has complex structures due to many joints, links, and axes. The spherical six-bar linkage shown in Figure 2 is designed using an origami-based pattern and fabricated through a layer-based method: smart composite microstructures (SCM). This process provides lightweight structures by replacing pin joints with flexures and finally enables us to achieve spherical six-bar linkage by folding a paper just three times.

Figure 3 shows the pattern for the origami-based spherical six-bar linkage. All axes of the flexures pass through a single point and rotate around the point. The spherical six-bar array is installed between the wing and the crank as shown in Figure 2. The spherical linkage acts as a mechanical filter, which transmits only linear motion to the wings. This is possible since one end of the linkage is constrained. The motor and the cranks make the rotating motion. When this rotating motion passes through the spherical linkage, the horizontal motion is absorbed by the spherical linkage but the vertical motion is transmitted to the wing.

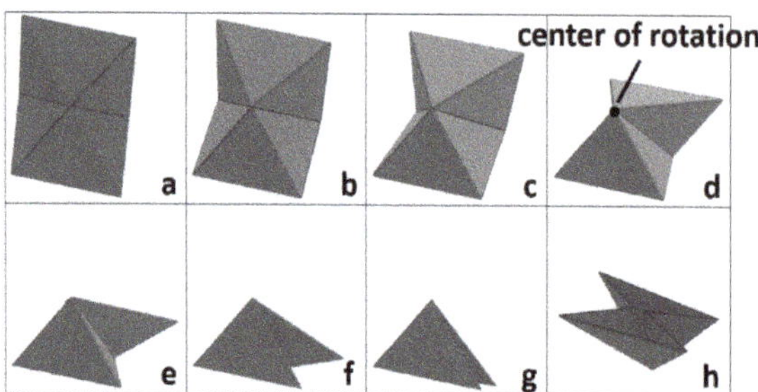

Figure 3. Snapshots of the origami-based spherical six-bar linkage. All axes of the flexure joints (folding line) pass through the rotation center indicated in (**d**). The pattern is folded from (**a–g**). The backside of the pattern is shown in (**h**).

Depending on how we choose to constrain the direction of the axis, the direction of the output motion can be differed. In our case, we need a flapping motion vertical to the motor's axis. Therefore, the axis constrained has been determined as a direction perpendicular to the span wise direction, which is equal to the direction of the motor's axis. Figure 2 shows how the constrained spherical six-bar works.

2.2. Fabrication

The proposed mechanism has been designed to have wingspan of 125 mm, which is referred to as the size of a beetle [32,33]. The beetle has appropriate size and mass, considering our fabrication capacity—the machining resolution is about 500 µm and this resolution is not enough to make micro-scale flappers. The body has the size of 20 mm × 10 mm × 10 mm and the whole mechanism weighs 5.7 g. Its planar design is shown in Figure 4B, which consist of twenty joints and eighteen links. To manufacture and assemble these tiny and complex kinematic structures using conventional pins and links needs much effort and the structure tends to be heavy because of the increased number of components.

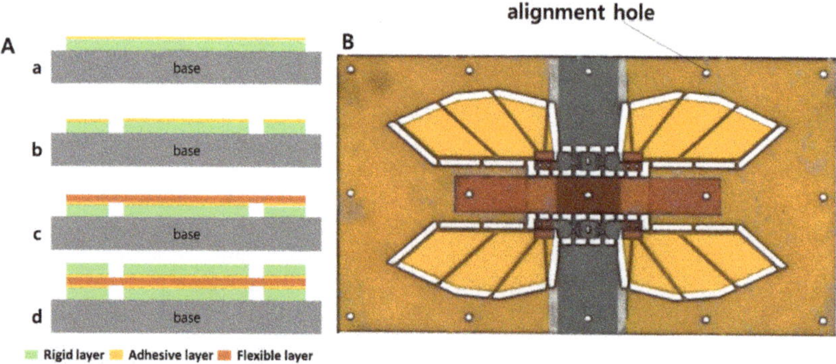

Figure 4. (**A**) The process of smart composite microstructures (SCM). The rigid layer with the adhesive is prepared (**a**). The joints are selectively cut by the laser machine (**b**). The flexible layer (rip-stop fabric) is sandwiched between two rigid layers (**c,d**). (**B**) The manufactured flapping mechanism before assembly. The mechanism is assembled by folding the flexible joints.

To avoid these issues, the SCM process is applied [30]. The SCM process replaces the conventional pin joints and links with the flexure joints and composite links. In Figure 4, the whole fabrication process is shown and described. For the proposed mechanism, a glass fiber board (200 μm thickness) is used as the rigid links and the rip-stop fabric (200 μm thickness) is employed as the flexure joints. The flapping wings are also designed to be manufactured through the SCM process with the body frame. The wing has the size of 50 mm × 20 mm and includes glass fiber supporters to prevent unwanted bending induced by air resistance during flapping, which is shown in Figure 4B. For the wing membrane, 25 μm polyimide film has been applied. Due to the SCM process, the whole mechanism weighs 5.7 g and the mass budget is shown in Table 2.

Table 2. Mass Budget.

Components	Quantity (ea.)	Mass (g)	Ratio to Overall Mass (%)
Flapping wing	2	0.12	4.21
Motor	1	3.9	68.42
Transmission	2	0.24	8.42
Body frame	1	1.08	18.95
Total	-	5.7	100.0

3. Mechanism Analysis

Kinematic modeling is done to determine lengths of links and accordingly get the desired performance—stable flapping posture and high thrust. To achieve the desired performance, two conditions must be satisfied. First, the flapping angle output should be symmetric to prevent pitch torque. Second, in order to maintain the angle of attack (AoA) and gain sufficient momentum, the upstroke and downstroke speed need to be as equal as possible. If the above two conditions are satisfied, the relation between the input crank angle and the output flapping angle has an ideal cosine curve as shown in Figure 5. In Figure 5, an optimized case and an arbitrary case are given. To satisfy the two required conditions, the flapping mechanism needs to be designed so that the flapping angle curve passes through the point *a*, *b*, *c* and *d*. To investigate the geometrical requirement to fulfill the conditions, kinematic modeling and analysis have been done.

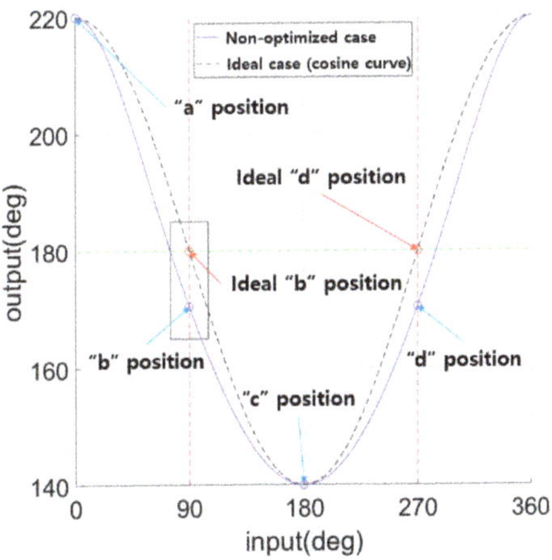

Figure 5. The secondary condition. A non-optimized case and an ideal case are simultaneously shown to compare the difference.

3.1. Kinematic Modeling

For three-dimensional analysis, the spherical coordinate system is employed as shown in Figure 6. Based on the coordinate, a vector loop formula is derived as follows:

$$\vec{r} + \vec{L_3} + \vec{L_2} + \vec{L_1} = \vec{G} \tag{1}$$

$$r \begin{bmatrix} \cos\theta_1 \\ \sin\theta_1 \\ 0 \end{bmatrix} + L_3 \begin{bmatrix} \cos\theta_2 \\ \sin\theta_2 \\ 0 \end{bmatrix} + L_2 \begin{bmatrix} \sin\phi_3\cos\theta_2 \\ \sin\phi_3\sin\theta_2 \\ \cos\phi_3 \end{bmatrix} + L_1 \begin{bmatrix} \sin\phi_4 \\ 0 \\ \cos\phi_4 \end{bmatrix} = \begin{bmatrix} G \\ 0 \\ -d \end{bmatrix} \tag{2}$$

where r, L_1, L_2 and L_3 are the length of the crank, the output link, the connecting rod, and the spherical bar, respectively. The angle parameters such as θ_1, ϕ_4, θ_2, and ϕ_3 are indicated in Figure 6.

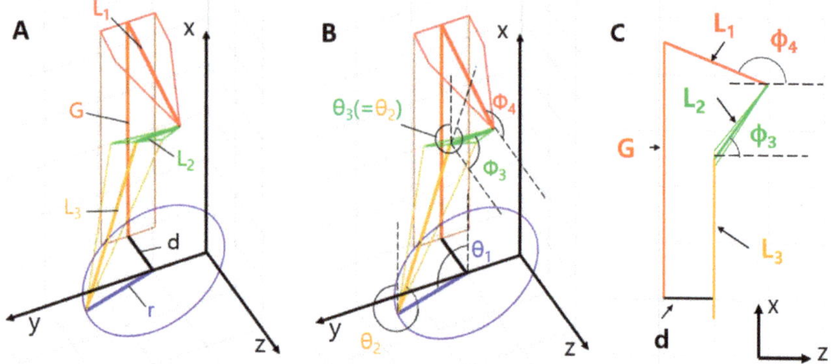

Figure 6. The vector loop of the constrained spherical six-bar transmission for the kinematic analysis. Diagonal view (**A**,**B**) and x-z plane view (**C**).

For the first condition, the transmission needs to be designed for the wings to have a symmetric flapping angle, $\pm 40°$, from the neutral position. When applied to this kinematic modeling, the highest and lowest values of ϕ_4 based on the neutral point ($\phi_4 = 180°$) are generated as $180° + 40°$ and $180° - 40°$, respectively.

$$r \begin{bmatrix} 1 \\ 0 \\ 0 \end{bmatrix} + L_3 \begin{bmatrix} 1 \\ 0 \\ 0 \end{bmatrix} + L_2 \begin{bmatrix} \sin\phi_3 \\ 0 \\ \cos\phi_3 \end{bmatrix} + L_1 \begin{bmatrix} \sin(180° + 40°) \\ 0 \\ \cos(180° + 40°) \end{bmatrix} = \begin{bmatrix} G \\ 0 \\ -d \end{bmatrix} \tag{3}$$

$$\begin{bmatrix} -1 \\ 0 \\ 0 \end{bmatrix} + L_3 \begin{bmatrix} 1 \\ 0 \\ 0 \end{bmatrix} + L_2 \begin{bmatrix} \sin\phi_3 \\ 0 \\ \cos\phi_3 \end{bmatrix} + L_1 \begin{bmatrix} \sin(180° - 40°) \\ 0 \\ \cos(180° - 40°) \end{bmatrix} = \begin{bmatrix} G \\ 0 \\ -d \end{bmatrix} \tag{4}$$

Based on the Equations (3) and (4), the relation between L_2 and L_3 is given as follows:

$$L_2{}^2 = (G - L_3)^2 + \left(\frac{r}{\tan 40°} - d \right)^2 \tag{5}$$

The secondary condition is to make the neutral of the crank and the neutral of the wing equal. In order to get the thrust during a wing-beat cycle, the key is to maintain the angle of attack (AoA) of the wing. Theoretically, when the equal shape and speed of the wings are given, $45°$ is the proper AoA for the maximum thrust [34].

In order to maintain the AoA, the wing angular velocities of upstroke and downstroke need to be as symmetrical and constant as possible. This condition is indicated in Figure 6, which shows the relation between the flapping angle and the crank angle. When the curve corresponds to the ideal cosine curve, the condition is satisfied. In Figure 6, the output flapping angle is π when the crank input angle is $\pi/2$ and $3\pi/2$. That is, it is preferable that the path length is the same as the path length between the highest point and the neutral point (from the point a and b) and the path length between the neutral point and the lowest point (from the point b to c).

$$r \begin{bmatrix} 0 \\ 1 \\ 0 \end{bmatrix} + L_3 \begin{bmatrix} \cos\theta_{2b} \\ \sin\theta_{2b} \\ 0 \end{bmatrix} + L_2 \begin{bmatrix} \sin\phi_{3b}\cos\theta_{2b} \\ \sin\phi_{3b}\sin\theta_{2b} \\ \cos\phi_{3b} \end{bmatrix} + L_1 \begin{bmatrix} 0 \\ 0 \\ -1 \end{bmatrix} = \begin{bmatrix} G \\ 0 \\ -d \end{bmatrix} \tag{6}$$

$$r \begin{bmatrix} 0 \\ -1 \\ 0 \end{bmatrix} + L_3 \begin{bmatrix} \cos\theta_{2d} \\ \sin\theta_{2d} \\ 0 \end{bmatrix} + L_2 \begin{bmatrix} \sin\phi_{3d}\cos\theta_{2d} \\ \sin\phi_{3d}\sin\theta_{2d} \\ \cos\phi_{3d} \end{bmatrix} + L_1 \begin{bmatrix} 0 \\ 0 \\ -1 \end{bmatrix} = \begin{bmatrix} G \\ 0 \\ -d \end{bmatrix} \tag{7}$$

where $\phi_{3b} = \phi_{3d}$, $\theta_{2b} + \theta_{2d} = 2\pi$. θ_{2b} and θ_{2d} are when θ_2 is in point b and d, respectively, as shown in Figure 6.

Based on the Equations (6) or (7), the relation between L_2 and L_3 is given as follows:

$$(L_3 + L_2 \sin\phi_{3b \text{ or } 3d})^2 = G^2 + r^2 \tag{8}$$

$$L_2 \cos\phi_{3b \text{ or } 3d} - L_1 = -d \tag{9}$$

The term, $\cos\phi_{3b \text{ or } 3d}$, is eliminated from the Equations (8) and (9). In result, following equation is given:

$$L_2 = \sqrt{\left(\sqrt{G^2 + r^2} - L_3\right)^2 + (L_1 - d)^2} \tag{10}$$

By using the two conditions—Equations (5) and (10)—optimized geometrical values can be determined.

3.2. Parameter Selection

Considering the whole size of the flapping mechanism, L_1, G, d and r are initially determined as 10.11 mm, 22.75 mm, 4.5 mm, and 6.5 mm, respectively. Figure 7a shows the change in the output flapping angle according to the length ratio of L_2 and L_3 to the crank length r.

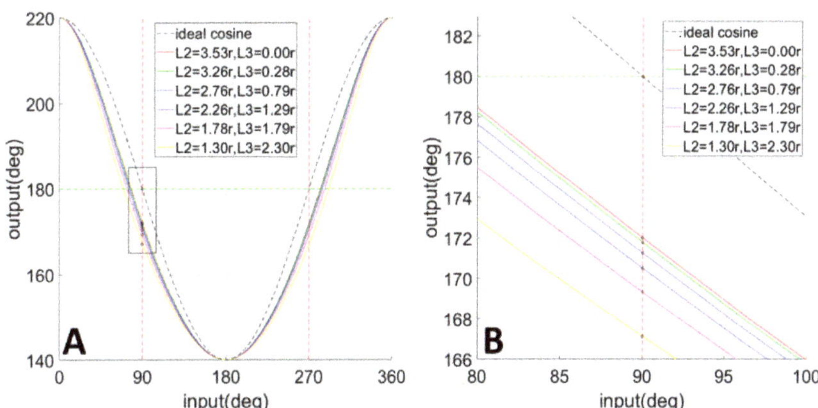

Figure 7. (**A**) Simulated output flapping angle vs. input crank angle by varying L_2 and L_3. (**B**) The magnification of the rectangle indicated in (**A**).

In Figure 7b, as L_2 increases and L_3 decreases, the curve becomes similar to the ideal cosine. The case when L_2 is 3.53 r and L_3 is 0.00 r is the closest one. However, it is difficult to make a certain link length zero because of realistic problems such as material thickness, or inter-part interference. Therefore, L_2 and L_3 are determined as 8.38 mm (=1.30 r) and 14.72 mm (=2.30 r), respectively, which correspond to the yellow line in Figure 7.

4. Experimental Results

To check whether the proposed transmission and the flapping mechanism properly works or not, experimental tests have been done. By using the high-speed camera, the flapping angle, frequency, and wing rotation angle are observed. In addition, the thrust is measured using a load cell to investigate the performance of the flapping mechanism.

4.1. Flapping Mechanism

Figures 8 and 9 show the snapshots during flapping. The snapshots are taken by a high-speed camera with the frame rate of 900. To investigate the flapping angle and the wing rotation angle, the mechanism is filmed from the top and side. To operate the flapping mechanism, 3.8 V and 580 mA are given to the DC motor. Based on the Figures 8 and 9 the mechanism operates with the flapping frequency of 23 Hz.

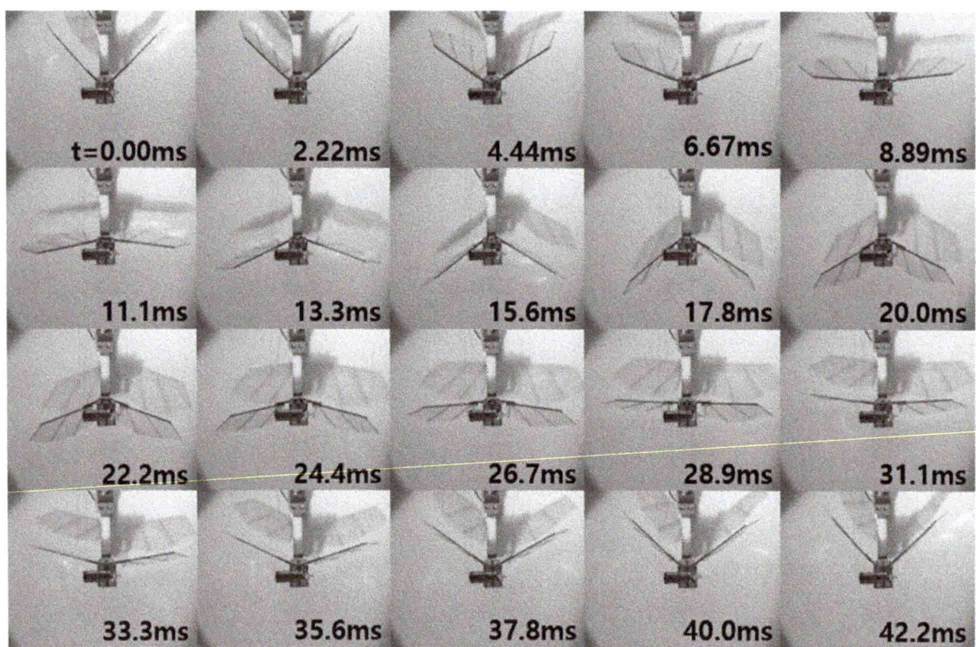

Figure 8. Snapshots while flapping at 23 Hz. A high-speed video of flapping, taken at 900 fps. Flapping angle can be measured from the top view.

The flapping angle is shown in Figure 8. For the right wing, the flapping angles are measured as +42° and −37° with respect to the neutral point; 50° and −36.5° are measured in case of the left wing. This result is confirmed through Figure 10. Figure 10 shows the flapping angles depending on the crank angle. The left wing has flapping angle range of 86.5° (141.5°~228°) and the right wing has 77° (144°~223°). There exists a 10° difference between two wings. The difference may occur due to the problems coming from fabrication and assembly issues.

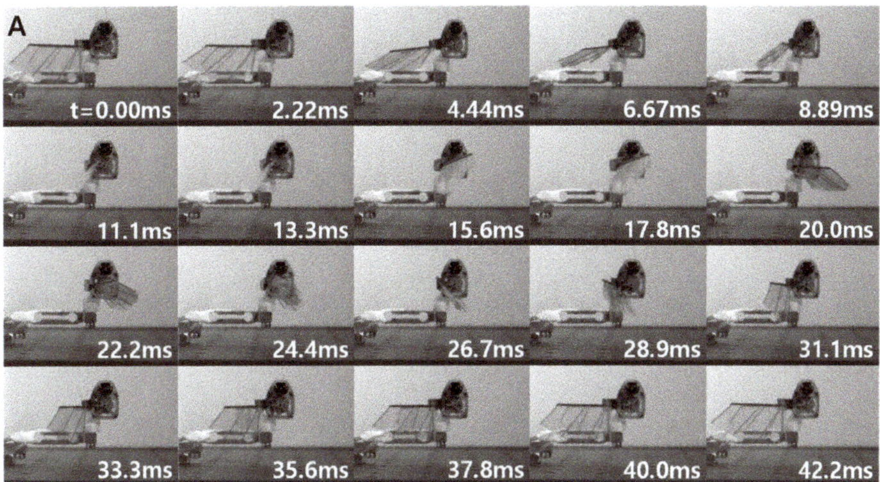

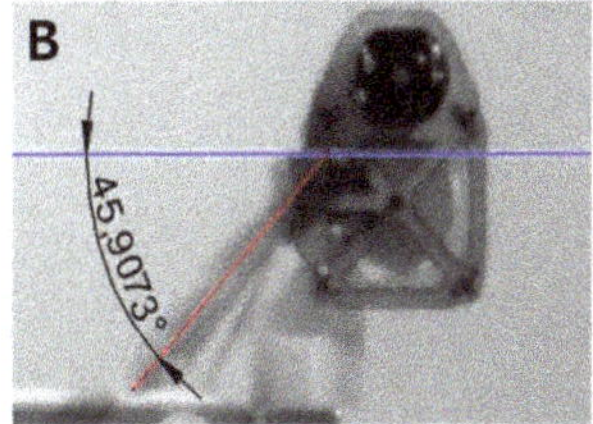

Figure 9. (**A**) Side view of the flapping mechanism at 23 Hz, taken at 900 fps. (**B**) Wing rotation angle can be measured.

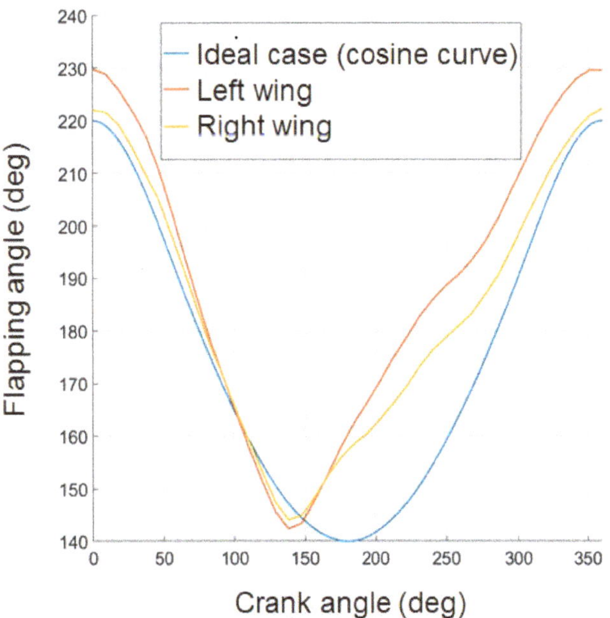

Figure 10. Comparison of flapping angle between the experimental case and the ideal case.

As previously said, in terms of overall trend, is it preferable that the relation between the flapping angle and the crank angle is close to the ideal cosine curve. However, Figure 10 shows the substantial difference between the experimental results and the ideal cosine curve. Especially for the upstroke, the gap between the experiment and the cosine curve tends to increase. In addition, the upstroke requires much time, compared to the downstroke. This can be explained by fabrication tolerance. This tolerance induces the flexures to make delay when they transmit the motion. During the upstroke, therefore, the motor's motion is transmitted with delay. However, in case of the downstroke, the facets structurally help to transmit the motor's motion and accordingly the delay does not occur.

4.2. Thrust

To check whether the proposed transmission properly generates the thrust or not, force measurement has been done. The data are collected through a load cell shown in Figure 11, sampled with the rate of 1000 Hz and filtered with the cut-off frequency of 100 Hz. The filtered data and the averaged data are given in Figure 12. The average thrust is measured as 2.34 gf (=22.9 mN) at the flapping frequency of 23 Hz.

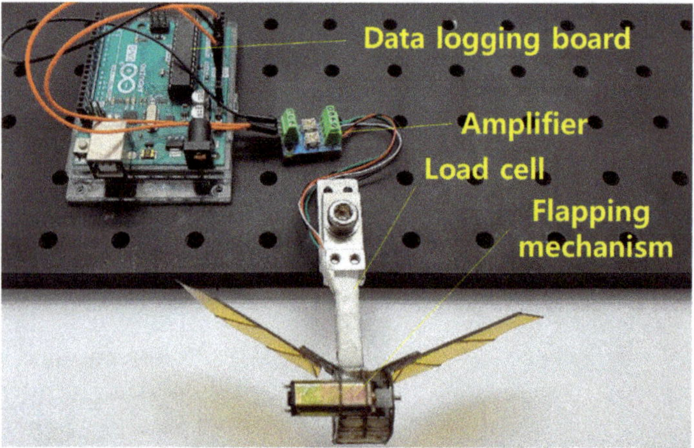

Figure 11. Experimental setup. A load cell has been used to measure the thrust. The signal goes to the amplifier and the amplified data are logged to the PC.

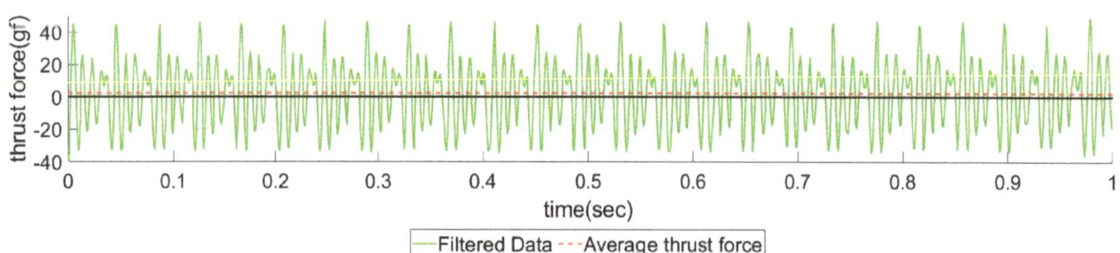

Figure 12. The measured thrust. The low-pass filtered data (solid line) and the average value (dashed line) are given.

The thrust is successfully generated since the wing rotation angle is observed as we intended, which is shown in Figure 11B. For the wing rotation, a passive hinge has been applied to get an additional degree of freedom. The stoppers, however, have been added to prevent the excessive wing rotation. As a result, an angle of attack (AoA) of about 45° has been observed as shown in Figure 11B. This AoA maintains during both the upstroke and downstroke. Consequently, the thrust can be generated.

5. Conclusions

In this paper, we proposed a novel transmission to reduce the assembly process and wear problems for DC motor-based FWMAVs. To make this possible, a geometrically constrained and origami-based spherical six-bar linkage has been suggested. The origami-based fabrication method reduces the relative moving parts such as pin joints and sliders, which resolves mechanical wear problems and assembly issues. The constrained spherical six-bar linkage passes only necessary motions by filtering the DC motor's rotating movement. Therefore, the linearly reciprocating movement is transferred and the flapping motion is successfully made. To demonstrate the feasibility of the proposed concept, flapping tests have been done to investigate the flapping angle, the wing rotation angle, and the vertical thrust. As a result, the suggested mechanism has shown a flapping angle of about 80°, a 45° wing rotation angle, 23 Hz of flapping frequency, and 2.34 gf (=22.9 mN) of the vertical thrust. Although the thrust is not enough to take off, we confirmed that the constrained spherical six-bar linkage successfully works as a transmission for the DC motor-based FWMAVs.

Supplementary Materials: The following are available online at https://www.mdpi.com/2076-341 7/11/4/1515/s1, Video S1: Working process of the constrained spherical six-bar.

Author Contributions: Conceptualization, S.-Y.B. and G.-P.J.; methodology, S.-Y.B. and G.-P.J.; software, S.-Y.B.; validation, S.-Y.B. and G.-P.J.; formal analysis, S.-Y.B.; investigation, S.-Y.B.; resources, G.-P.J. and J.-S.K.; data curation, S.-Y.B.; writing—original draft preparation, S.-Y.B., J.-S.K. and G.-P.J.; writing—review and editing, J.-S.K. and G.-P.J.; visualization, S.-Y.B.; supervision, G.-P.J.; project administration, G.-P.J.; funding acquisition, G.-P.J. All authors have read and agreed to the published version of the manuscript.

Funding: This research was supported by a grant to Bio-Mimetic Robot Research Center Funded by Defense Acquisition Program Administration, and by Agency for Defense Development (UD190018ID). This research was also supported by Korea Institute for Advancement of Technology (KIAT) grant funded by the Korea Government (MOTIE) (P0008473, HRD Program for Industrial Innovation).

Institutional Review Board Statement: Not applicable.

Informed Consent Statement: Not applicable.

Data Availability Statement: Not applicable.

Conflicts of Interest: The authors declare no conflict of interest.

References

1. Takahashi, H.; Abe, K.; Takahata, T.; Shimoyama, I. Experimental study of the aerodynamic interaction between the forewing and hindwing of a beetle-type ornithopter. *Aerospace* **2018**, *5*, 83. [CrossRef]
2. Karásek, M.; Muijres, F.T.; De Wagter, C.; Remes, B.D.; de Croon, G.C. A tailless aerial robotic flapper reveals that flies use torque coupling in rapid banked turns. *Science* **2018**, *361*, 1089–1094. [CrossRef] [PubMed]
3. Gong, D.; Lee, D.; Shin, S.J.; Kim, S. Design and Experiment of String-based Flapping Mechanism and Modulized Trailing Edge Control System for Insect-like FWMAV. In Proceedings of the 2018 AIAA Information Systems-AIAA Infotech@ Aerospace, Kissimmee, FL, USA, 8–12 January 2018; p. 0987.
4. De Wagter, C.; Karásek, M.; de Croon, G. Quad-thopter: Tailless flapping wing robot with four pairs of wings. *Int. J. Micro Air Veh.* **2018**, *10*, 244–253. [CrossRef]
5. Oppenheimer, M.W.; Sigthorsson, D.; Doman, D.B.; Weintraub, I. Wing design and testing for a tailless flapping wing micro-air vehicle. In Proceedings of the AIAA Guidance, Navigation, and Control Conference, Grapevine, TX, USA, 9–13 January 2017; p. 1271.
6. Yan, X.; Qi, M.; Lin, L. Self-lifting artificial insect wings via electrostatic flapping actuators. In Proceedings of the 2015 28th IEEE International Conference on Micro Electro Mechanical Systems (MEMS), Estoril, Portugal, 18–22 January 2015; pp. 22–25.
7. Hines, L.; Colmenares, D.; Sitti, M. Platform design and tethered flight of a motor-driven flapping-wing system. In Proceedings of the 2015 IEEE International Conference on Robotics and Automation (ICRA), Seattle, DC, USA, 26–30 May 2015; pp. 5838–5845.
8. Ristroph, L.; Childress, S. Stable hovering of a jellyfish-like flying machine. *J. R. Soc. Interface* **2014**, *11*, 20130992. [CrossRef] [PubMed]
9. Lau, G.-K.; Chin, Y.-W.; Goh, J.T.-W.; Wood, R.J. Dipteran-insect-inspired thoracic mechanism with nonlinear stiffness to save inertial power of flapping-wing flight. *IEEE Trans. Robot.* **2014**, *30*, 1187–1197. [CrossRef]

10. Hines, L.; Campolo, D.; Sitti, M. Liftoff of a motor-driven, flapping-wing microaerial vehicle capable of resonance. *IEEE Trans. Robot.* **2013**, *30*, 220–232. [CrossRef]

11. Gaissert, N.; Mugrauer, R.; Mugrauer, G.; Jebens, A.; Jebens, K.; Knubben, E.M. Inventing a micro aerial vehicle inspired by the mechanics of dragonfly flight. In Proceedings of the Conference Towards Autonomous Robotic Systems, London, UK, 3–5 July 2019; pp. 90–100.

12. Sahai, R.; Galloway, K.C.; Wood, R.J. Elastic Element Integration for Improved Flapping-Wing Micro Air Vehicle Performance. *IEEE Trans. Robot.* **2013**, *29*, 32–41. [CrossRef]

13. Keennon, M.; Klingebiel, K.; Won, H. Development of the nano hummingbird: A tailless flapping wing micro air vehicle. In Proceedings of the 50th AIAA Aerospace Sciences Meeting Including the New Horizons Forum and Aerospace Exposition, Nashville, TN, USA, 9–12 January 2012; p. 588.

14. Campolo, D.; Azhar, M.; Lau, G.-K.; Sitti, M. Can DC motors directly drive flapping wings at high frequency and large wing strokes? *IEEE/ASME Trans. Mechatron.* **2012**, *19*, 109–120. [CrossRef]

15. Galiński, C.; Żbikowski, R. Insect-like flapping wing mechanism based on a double spherical Scotch yoke. *J. R. Soc. Interface* **2005**, *2*, 223–235. [CrossRef]

16. Roshanbin, A.; Altartouri, H.; Karasek, M.; Preumont, A. COLIBRI: A hovering flapping twin-wing robot. *Int. J. Micro Air Veh.* **2017**, *9*, 270–282. [CrossRef]

17. Phan, H.V.; Au, T.K.L.; Park, H.C. Clap-and-fling mechanism in a hovering insect-like two-winged flapping-wing micro air vehicle. *R. Soc. Open Sci.* **2016**, *3*, 160746. [CrossRef]

18. Phan, H.V.; Truong, Q.T.; Au, T.K.L.; Park, H.C. Optimal flapping wing for maximum vertical aerodynamic force in hover: Twisted or flat? *Bioinspiration Biomim.* **2016**, *11*, 046007. [CrossRef] [PubMed]

19. Phan, H.V.; Kang, T.; Park, H.C. Design and stable flight of a 21 g insect-like tailless flapping wing micro air vehicle with angular rates feedback control. *Bioinspiration Biomim.* **2017**, *12*, 036006. [CrossRef]

20. Wood, R.J. The First Takeoff of a Biologically Inspired At-Scale Robotic Insect. *IEEE Trans. Robot.* **2008**, *24*, 341–347. [CrossRef]

21. Fuller, S.B. Four wings: An insect-sized aerial robot with steering ability and payload capacity for autonomy. *IEEE Robot. Autom. Lett.* **2019**, *4*, 570–577. [CrossRef]

22. Arabagi, V.; Hines, L.; Sitti, M. Design and manufacturing of a controllable miniature flapping wing robotic platform. *Int. J. Robot. Res.* **2012**, *31*, 785–800. [CrossRef]

23. Steltz, E.; Wood, R.J.; Avadhanula, S.; Fearing, R.S. Characterization of the micromechanical flying insect by optical position sensing. In Proceedings of the 2005 IEEE International Conference on Robotics and Automation, Barcelona, Spain, 18–22 April 2005; pp. 1252–1257.

24. Roll, J.A.; Cheng, B.; Deng, X. Design, fabrication, and experiments of an electromagnetic actuator for flapping wing micro air vehicles. In Proceedings of the 2013 IEEE International Conference on Robotics and Automation, Karlsruhe, Germany, 6–10 May 2013; pp. 809–815.

25. Roll, J.A.; Cheng, B.; Deng, X. An electromagnetic actuator for high-frequency flapping-wing microair vehicles. *IEEE Trans. Robot.* **2015**, *31*, 400–414. [CrossRef]

26. Bontemps, A.; Vanneste, T.; Paquet, J.; Dietsch, T.; Grondel, S.; Cattan, E. Design and performance of an insect-inspired nano air vehicle. *Smart Mater. Struct.* **2012**, *22*, 014008. [CrossRef]

27. Meng, K.; Zhang, W.; Chen, W.; Li, H.; Chi, P.; Zou, C.; Wu, X.; Cui, F.; Liu, W.; Chen, J. The design and micromachining of an electromagnetic MEMS flapping-wing micro air vehicle. *Microsyst. Technol.* **2012**, *18*, 127–136. [CrossRef]

28. Zou, Y.; Zhang, W.; Zhang, Z. Liftoff of an electromagnetically driven insect-inspired flapping-wing robot. *IEEE Trans. Robot.* **2016**, *32*, 1285–1289. [CrossRef]

29. Suzuki, K.; Shimoyama, I.; Miura, H. Insect-model based microrobot with elastic hinges. *J. Microelectromech. Syst.* **1994**, *3*, 4–9. [CrossRef]

30. Wood, R.J.; Avadhanula, S.; Sahai, R.; Steltz, E.; Fearing, R.S. Microrobot design using fiber reinforced composites. *J. Mech. Des.* **2008**, *130*, 052304. [CrossRef]

31. Koh, J.-S.; Cho, K.-J. Omega-Shaped Inchworm-Inspired Crawling Robot with Large-Index-and-Pitch (LIP) SMA Spring Actuators. *IEEE/ASME Trans. Mechatron.* **2013**, *18*, 419–429. [CrossRef]

32. Nguyen, Q.V.; Park, H.C.; Goo, N.S.; Byun, D. Characteristics of a Beetle's Free Flight and a Flapping-Wing System that Mimics Beetle Flight. *J. Bionic Eng.* **2010**, *7*, 77–86. [CrossRef]

33. Nguyen, Q.V.; Truong, Q.T.; Park, H.C.; Goo, N.S.; Byun, D. A motor-driven flapping-wing system mimicking beetle flight. In Proceedings of the 2009 IEEE International Conference on Robotics and Biomimetics (ROBIO), Bangkok, Thailand, 22–25 February 2009; pp. 1087–1092.

34. Steltz, E.; Avadhanula, S.; Fearing, R.S. High lift force with 275 Hz wing beat in MFI. In Proceedings of the 2007 IEEE/RSJ International Conference on Intelligent Robots and Systems, San Diego, CA, USA, 29 October–2 November 2007; pp. 3987–3992.

Article

Bioinspired Divide-and-Conquer Design Methodology for a Multifunctional Contour of a Curved Lever

Jehyeok Kim, Junyoung Moon, Jaewook Ryu and Giuk Lee *

School of Mechanical Engineering, Chung-Ang University, Seoul 06974, Korea; michaeljhkim@cau.ac.kr (J.K.); mjyoung5@cau.ac.kr (J.M.); wodnr1958@cau.ac.kr (J.R.)
* Correspondence: giuklee@cau.ac.kr

Abstract: In this study, we propose a bioinspired design methodology for a multifunctional lever based on the morphological principle of the lever mechanism in the *Salvia pratensis* flower. The proposed divide-and-conquer contour design methodology does not treat a lever contour as a single curve that satisfies multiple functions. Rather, the lever contour combines partial contours to achieve its assigned subfunction. This approach can simplify the complex multifunctional problem in lever design. We include a case study of a lever utilized in a compact variable gravity compensator (CVGC) to explain the methodology in more detail. In the case study, four partial contours were designed to satisfy three types of functional requirements. The final design for the lever contour was manufactured and verified with visual measurement experiments. The experimental result shows that each partial contour successfully achieved its subfunctions.

Keywords: curved lever; lever design methodology; variable pivot of lever

Citation: Kim, J.; Moon, J.; Ryu, J.; Lee, G. Bioinspired Divide-and-Conquer Design Methodology for a Multifunctional Contour of a Curved Lever. *Appl. Sci.* **2021**, *11*, 6015. https://doi.org/ 10.3390/app11136015

Academic Editors: TaeWon Seo, Dongwon Yun and Gwang-Pil Jung

Received: 28 May 2021
Accepted: 25 June 2021
Published: 28 June 2021

Publisher's Note: MDPI stays neutral with regard to jurisdictional claims in published maps and institutional affiliations.

1. Introduction

Lever mechanisms have been used to amplify force or displacement for a mechanical advantage since time immemorial. Based on the conservation of energy, the amount of amplification a lever provides can be calculated as the ratio of the distance from the pivot point to the effort point and that from the pivot point to the load point. Numerous mechanical mechanisms and devices have been developed successfully with proper lever mechanisms [1–6]. H. Lin and W. Feng proposed a torque generator-driven lever mechanism for a torsional mirror. In the mechanism, a lever was exploited to enlarge traveling distance in limited design space [1]. Woude et al. developed a new design for a lever-propelled wheelchair [2]. K. Li and M. Gohnert suggested a lever mechanism for improving vibration isolation performance. In their study, damping force was conveyed to the mass through the lever [3]. Several researchers have developed gripper mechanisms for microelectromechanical systems (MEMS) using lever mechanisms. C. Shi et al. proposed a microgripper with a large magnification ratio. In this mechanism, the lever successfully amplified the limited output displacement of a lead zirconate titanate (PZT) actuator for the desired displacement of the gripper [4]. K. Kwon et al. developed a linear motor using a lever mechanism to improve the position accuracy. Unlike this mechanism of a microgripper, the lever mechanism in a linear motor had the role of reducing the displacement [5]. M. French and M. Widden developed a static balancing mechanism using a spring and lever. In this mechanism, the lever amplified the spring force and achieved a lightweight static balancer for a lamp [6].

In recent years, lever mechanisms with variable amplification ratios have frequently been applied in robot mechanisms, particularly variable stiffness actuators [7–12]. A. Jafari et al. [7] and N. Tsagarakis [8] et al. proposed adjustable stiffness actuators with class 1 levers for ankle assistance mechanisms. Sun et al. [10], Groothuis et al. [11] and Barrett et al. [12] also developed a class 1 lever-based actuator to achieve variable stiffness. To adjust the stiffness, the lever position was varied using an additional motor. B. Kim and

J. Song also developed a variable stiffness actuator with a lever mechanism. When the class 2 lever mechanism was applied, the position of the lever's load point varied to change the actuator's stiffness [9]. Although the lever mechanisms with a variable amplification ratio achieved the desired function successfully, the shape for the levers was a straight line. Moreover, due to the simplicity of the straight lever, the above-mentioned studies did not propose a lever contour design methodology.

However, when a lever had an additional required function beyond amplification such as preventing intervention between design spaces or maximizing the variable range of amplification ratio, the lever was required to have a shape other than a simple straight line. J. Kim et al. developed a compact variable gravity compensation mechanism with a non-straight lever shape. An angled lever was devised to achieve the amplification of the spring force and also to satisfy the circular design space, [13]. M. Dezman and A. Gams suggested an arc-shaped lever-based compliant actuator to satisfy both force amplification and energy-efficient stiffness variation functions [14]. Although these studies showed a non-straight lever shape, the geometries of the levers' contours were still simple and did not require any particular contour design methodology.

J. Kim et al. [15] proposed a curved lever design to improve the performance of a variable gravity compensation mechanism. In this method, the curved lever was mathematically modeled based on B-spline representation and optimized using a genetic algorithm. Although the research suggested a curved lever design methodology, there was a limitation that the lever contour was represented as a unitary curve. This approach is suitable for a nonfunctional lever contour. However, when a multifunctional lever contour is designed using a unitary curve model and optimization approach, the complexity of the design problem could worsen due to the difficulties of determining the design vector and constructing a multivariable objective function with weighting factors.

Recently, we proposed a new version of a compact variable gravity compensator with a multifunctional lever [16]. In this study, first, we proposed the early concept of divide-and-conquer design methodology for a curved lever contour. Although the lever contour, on the paper, showed the potential of the divide-and-conquer design methodology, one of the lever contours still had a curve representation for two assigned functions. When the lever contour was optimized for two variable objective functions, design variables that maximize each function could not be selected.

To overcome the limitation of the above-mentioned lever contour design methods, in this study, we proposed a bioinspired divide-and-conquer design methodology for multifunctional lever contours. *Salvia pratensis* flowers have a curved lever mechanism in the staminal lobe as an effective pollen dispensing system. The staminal lever curve can be subdivided into a contour to prevent interference with bees' heads and a contour for applying pollen to bees' abdomens. From this morphological insight, we subdivided all required functions into subfunctions that were assigned to the design goal of each partial contour. Each contour was represented by a mathematical curve model suitable for achieving the assigned function and was determined through an appropriate method. The design methodology provides a design tool for the designer to generate a multifunctional lever contour.

To convey the usage of the methodology, we determined the compact variable gravity compensation (CVGC) mechanism as a design case study. Four subfunctions were assigned to the curved lever, which included: (1) minimizing the required force for the torque variation, (2) maximizing the variable range of the compensation torque, (3) preventing interference with cam structure, and (4) connecting between contours with C1 continuity. The experiments showed that the derived lever contour successfully achieved all subfunctions simultaneously. Moreover, the lever contour showed a 1.3 times wider variable range as compared with the lever in [16]. This paper is organized as follows: In Section 2, we explain the bioinspired insight from *Salvia pratensis* flowers and the concept of the lever contour design methodology in detail; in Section 3, we describe the operating concept of CVGC and the required functions for the curve lever mechanism; in Section 4, we discuss the detailed design processes for the lever contour; in Section 5, we show the experimental results of the designed curved lever contour; and finally, in Section 6, we present our conclusions.

2. Bioinspired Curved Lever Contour Design Methodology

Divide-and-conquer design methodology for the multifunctional contour of the curved lever was developed, inspired by a staminal lever system of the *Salvia pratensis* flower. To convey the bioinspired features of the methodology, the morphological characteristics of the staminal lever are illustrated in Section 2.1. In Section 2.2, the design methodology is explained in detail.

2.1. Curved Lever Mechanism in Salvia pratensis Flower

The *Salvia pratensis* flower has a staminal lever system that can dispense pollen to flower-visiting bees' abdomens for successful reproduction [17]. As shown in Figure 1a, the lever system is composed of a connective plate, a fertile anther lobe, and a rotational joint that has the role of a pivot. When a bee pushes its proboscis into the flower, the connective plate is forced to the left, as in Figure 1a. This force rotates the connective plate and fertile anther lobe clockwise. Therefore, the theca placed at the end of anther lobe makes contact with the bee's abdomen.

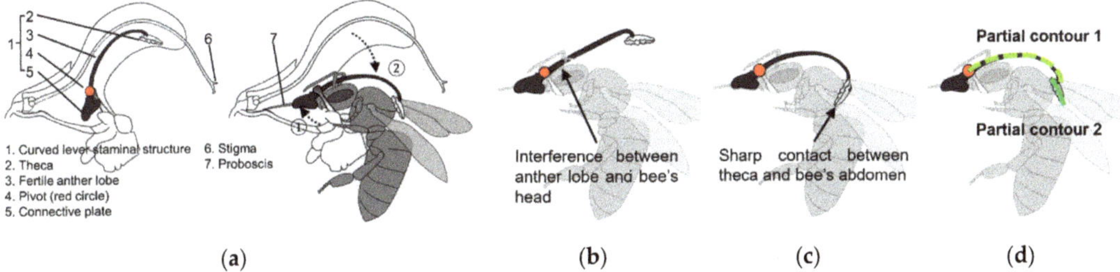

Figure 1. Schematic drawing of the lever mechanism in a *Salvia pratensis* flower and its interaction with a flower-visiting bee as depicted by Reith et al. [17]: (**a**) Longitudinal section of a flower and motions of the curved staminal lever; (**b**) a hypothetical straight-line staminal lever; (**c**) a hypothetical staminal lever where the theca has the same curvature as the anther lobe; (**d**) a real staminal lever composed of two partial contours. The partial contours 1 and 2 prevent interference with the bee's head and avoid sharp contact between the theca and the bee's abdomen, respectively.

The contour of the lever has a proper curved shape to achieve the successful dispensing of pollen. As shown in Figure 1b, if the contour of an anther lobe was a straight line, the theca could not make contact with a bee's abdomen due to interference with the bee's head. In addition, if the contour of the theca had the same curvature as the anther lobe shown in Figure 1c, the theca would make sharp contact with the abdomen and fail to dispense pollen effectively due to the narrow contact area.

As illustrated in Figure 1d, the entire lever curve can be seen as a combination of partial contours with individual functional goals. The partial Contour 1 has a proper curve to satisfy the function of preventing interference with the bee's head. The partial contour 2 also has a shape that achieves the function of maximizing the contact area between the theca and the bee's abdomen.

2.2. Concept of Bioinspired Design Methodology

As shown in Figure 2, the divide-and-conquer contour design methodology for a multifunctional lever was developed using bioinspired insights. This methodology fundamentally does not treat a lever contour as a single curve that should satisfy multiple functions. Rather, a lever contour is regarded as a combination of partial contours responsible for each subfunction. The procedure for designing a lever with this methodology is as follows: At first, subfunctions should be generated by dividing the overall function. Second, the design space of the lever is also divided into subareas that affect each subfunction. Third, a partial contour is derived by selecting an appropriate curve expression method and design method in consideration of the assigned subfunctions. Finally, one lever is determined by combining the derived partial contours. The partial contours only consider achieving their own functional goal. Therefore, in this design methodology, the designer can choose the most appropriate mathematical model to represent each partial contour and proper determination method for the design variables of each model. The divide-and-conquer lever design methodology has several advantages as follows: First, it can simplify complex multifunctional design problems.

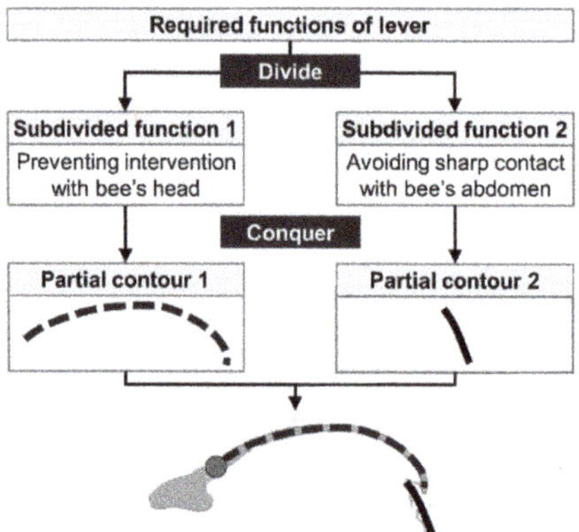

Figure 2. Morphological principle of the staminal curved lever of Salvia pratensis flower. The lever's required functions are subdivided into functions (divide) and assigned to each partial contour as design goals. Each partial contour has a proper shape only to satisfy their subdivided function (conquer). The combination of each partial contour forms the final contour of the curved lever.

A designer can concentrate on designing each partial contour to satisfy relatively simple functional goals. Next, this methodology makes it possible to construct efficient design processes in terms of mathematical modeling and the parameter determination method for each contour. Although some lever design problems may not suit this methodology because the multifunctionality cannot be divided well, the methodology can give a view to simplify the design problem.

3. Curved Lever Contour Design

To explain the methodology in more detail, a design case study was conducted for the actual mechanism of CVGC. As the lever mechanism in CVGC should achieve multiple functions, it is an appropriate case to show detailed processes. Before explaining the specific design processes, in this section, we briefly present the operating concept of the CVGC mechanism and the required functions of the lever.

3.1. Operating Concept of the CVGC Mechanism

As a variable gravity compensation mechanism, CVGC could generate compensation torque and change its amplitude to deal with variable gravitational torque. As shown in Figure 3, the core elements of the CVGC mechanism are the cam, cam follower, lever with a movable pivot, spring follower, and compression spring.

Figure 3a illustrates the principle of generating compensation torque. The rotation of an external load, which is the target mass to be compensated, leads to the rotation of the cam. The rotational motion of the cam makes the cam follower move along the linear guide and the movement of the cam follower rotates the lever clockwise about the pivot axis. This rotation of the lever pushes the spring follower to the right and compresses the spring. Consequently, the restoration force of the spring compresses the cam follower to the cam surface. Finally, due to the noncircular cam, the force at the cam surface generates a counterclockwise compensation torque at the cam.

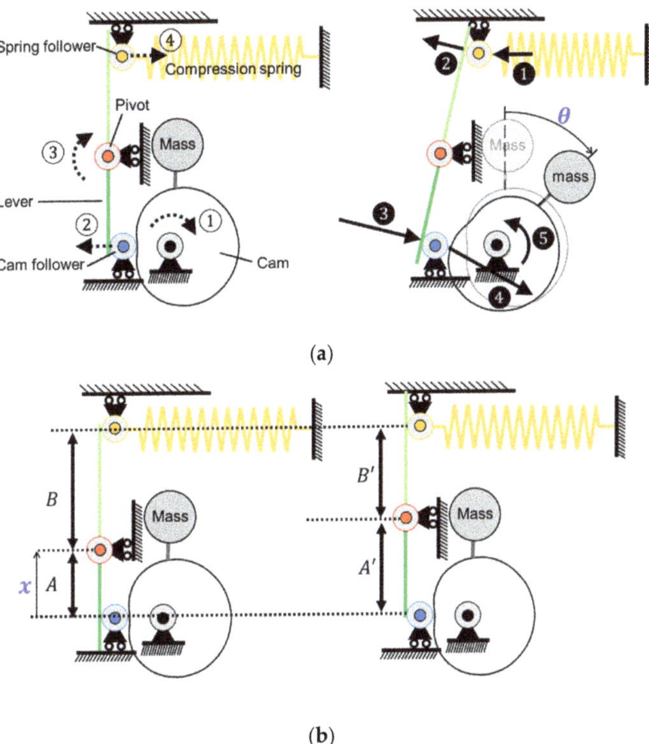

Figure 3. Operating concepts of CVGC for compensating a gravitational torque by mass: (**a**) Configuration and motions of CVGC. The white circled digits indicate the orders of motions describing the spring deformation due to the cam rotation. The black circled digits represent the order of forces that account for the compensating torque in the cam generated by the spring force. θ is the rotation angle of the cam; (**b**) principle of torque variation. The movement of the pivot can change the amplification ratio (B/A or B′/A′). B/A is larger than B′/A′, so it generates a larger compensation torque at the cam axis, and x denotes the distance between pivot position and cam follower.

Figure 3b illustrates the principle of compensation torque variation. The lever mechanism in CVGC can change its pivot position and the amplification ratio. Changing the ratio makes the magnitude of the force and the compensation torque generated on the cam controllable.

3.2. Three Types of Functional Requirements of Levers for CVGC

According to the abovementioned principles, the CVGC mechanism has been developed to achieve a compact and circular shape as shown in Figure 4. The lever has the role of amplifying the spring force and transmitting it to the cam in the limited design space. To maximize the performance of CVGC, the multiple functions that the lever should satisfy are: (1) minimizing the required force of the torque variation, (2) maximizing the variable range of the compensation torque, and (3) avoiding interference between the lever and other mechanical elements.

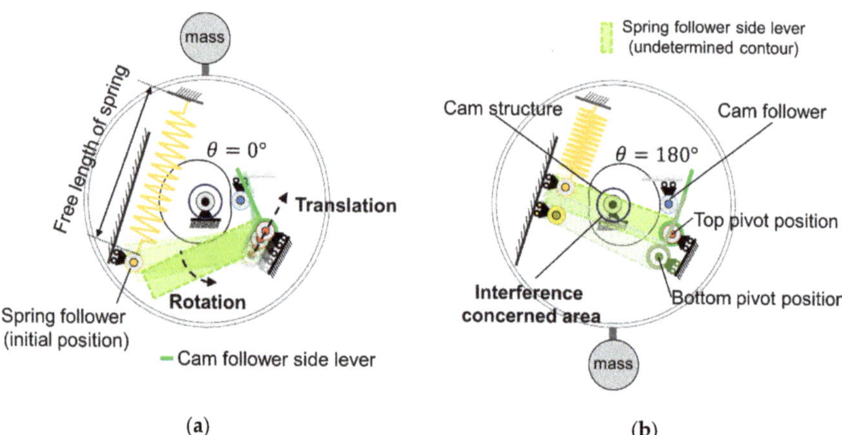

(a)   (b)

Figure 4. Functional requirements of the lever for CVGC: (**a**) The translation and rotation of the lever when the pivot position changes; (**b**) the maximum compressions of the spring at each top pivot position and bottom pivot position. Interference can occur between the lever and cam structure.

For the first function, minimizing the required force for the torque variation is essential to achieve high energy efficiency in the variable mechanism. When the variation is automated by an actuator, a large force is required which increases the actuator's size and weight and decreases the energy efficiency. Since the fundamental goal of CVGC is to create a compact mechanism, the first function should be satisfied by the lever design. The required force to change the pivot position can be calculated by the Lagrangian mechanics as follows:

$$F_{variation}(x) = -\frac{\partial PE_{spring}(x)}{\partial x} \tag{1}$$

x, $PE_{spring}(x)$, and $F_{variation}(x)$ denote the generalized coordinates, the force, and potential energy profile of the mechanism, respectively.

According to (1), if the energy variation is zero with respect to the change of pivot position, the required force can be zero in ideal frictionless conditions. The amount of spring deformation can be determined by the position of the spring follower. As shown in Figure 4, the rotation and translation of the lever occur simultaneously when the pivot changes. From the lever design perspective, the lever contour should not change the position of the spring follower when the pivot position varies.

Next, the second function of the lever is important because a wide variable range of compensation torque obviously allows CVGC to compensate for variable weight. The variable range of the compensation torque is calculated as the difference between the highest and lowest compensation torques. As illustrated in Figure 4b, for the CVGC mechanism that varies the compensation torque by changing the pivot position, it generates the highest torque when the pivot is placed at the top position of the pivot guide and generates the lowest torque when the pivot is placed at the bottom position.

According to [15], the maximum deformations of the spring at each pivot position of the CVGC mechanism could determine the amount of compensation torque. Therefore, to maximize the variable range of the torque, the lever contour should maximize the spring's deformation at the top pivot position as this is directly related to the highest torque. At the same time, this contour should minimize the spring's deformation at the bottom pivot position to minimize the lowest compensation torque. The design point to be considered in this process is that this lever contour must not compress the spring beyond its allowable deformation length.

As a final function of a lever, the lever contour should be designed to prevent interference with other mechanical elements. In particular, the cam structure, including the cam, camshaft, and bearings, is the main obstacle that hinders the lever from pressing the spring follower sufficiently. For the CVGC mechanism, the worst condition causing an interference problem between the lever contour and the cam structure is when the cam rotates 180° at the top pivot position. Therefore, as shown in Figure 4b, designing the lever contour to avoid interference at this worst condition is sufficient to prevent interference at other pivot positions.

4. Curved Lever Contour Design

To design the lever contour to achieve the multifunctionality mentioned in Section 3.2, the divide-and-conquer design methodology is applied. Through this method, each sub-function is assigned to the design goal of each partial contour. Partial contours are separately designed to satisfy their individual goals, and then combined as a final lever contour.

The geometrical design parameters of other mechanical components such as the cam profile, cam follower guides, and spring follower are given as geometrical constraints that those should not be able to pass through. The shape of the lever can be divided into the cam follower side and spring follower based on the pivot. In this section, it is assumed that the lever contour of the cam follower side is predetermined as a straight line and aims to design only the contour on the spring follower side.

4.1. Partial Contour 1 (PC1): Minimizing the Required Force for Torque Variation

The PC1 should not change the position of the spring follower when the pivot position varies along the pivot guide. The CVGC varies the compensation torque when the rotation angle of the cam is zero. Since the cam follower side of the lever always contacts the cam follower, the translation and rotation of the lever occur simultaneously when the pivot position changes, as illustrated in Figure 5a.

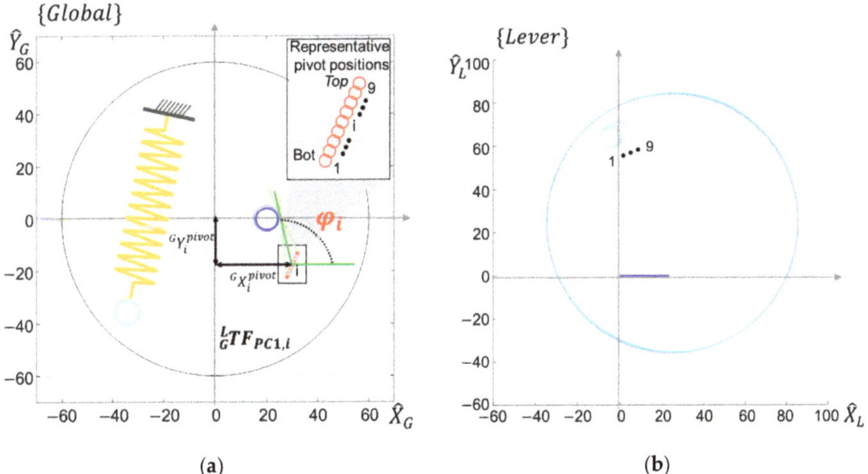

(a)　　　　　　　　　　(b)

Figure 5. *Cont.*

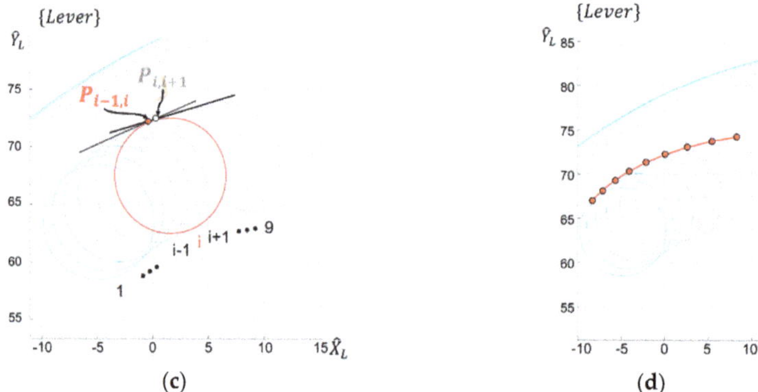

(c)

(d)

Figure 5. Design processes of the PC1. The units of x- and y-axes are mm: (**a**) Parameters to calculate transformation matrices at each representative pivot position; (**b**) spring followers are represented in the lever coordinate system; (**c**) common circumferential lines and points and principle to choosing the representative contact points; (**d**) the representative contact points of each spring follower and Hermite spline-based PC1.

In a global coordinate system, the contour design is complex due to the changes of relative angles and distances between the lever and the spring follower. However, the positions of points that determine the rigid lever shape do not change in the lever coordinate system. Therefore, the lever design problem can become easier when in a lever coordinate system.

To calculate the coordinate transformation, representative pivot positions are determined by dividing the pivot guidelines equally into n sections. Nine representative pivot positions were selected, including the top and bottom pivot positions. The unit length for a section and the rotation angles at each pivot position can be determined using (2) and (3) as follows:

$$D_x = \frac{{}^G X_9^{pivot} - {}^G X_1^{pivot}}{8}, \; D_y = \frac{{}^G Y_9^{pivot} - {}^G Y_1^{pivot}}{8} \tag{2}$$

$$\varphi_i = \frac{\pi}{2} + \sin^{-1}\frac{r_{cam}}{B_i} \tag{3}$$

where ${}^G X_i^{pivot}$ and ${}^G Y_i^{pivot}$ indicate the position of the ith pivot in the global coordinates; D_x and D_y are the x and y components of the unit length, respectively; r_{cam} is the radius of the cam follower. Then, the translation and transformation matrices at the ith selected pivot positions can be calculated using (4) and (5), respectively, as follows:

$$\begin{bmatrix} {}^G X_i^{pivot} \\ {}^G Y_i^{pivot} \end{bmatrix} = \begin{bmatrix} {}^G X_1^{pivot} + (i-1)D_x \\ {}^G Y_1^{pivot} + (i-1)D_y \end{bmatrix} \tag{4}$$

$${}^L_G TF_{PC1,i} = \begin{bmatrix} C_{-\varphi_i} & -S_{-\varphi_i} & 0 & -{}^G X_i^{pivot} \\ S_{-\varphi_i} & C_{-\varphi_i} & 0 & -{}^G Y_i^{pivot} \\ 0 & 0 & 1 & 1 \\ 0 & 0 & 0 & 1 \end{bmatrix} \tag{5}$$

where the left subscript G in (4) is the lever and global coordinate system, the left subscript G and the superscript L represent the global coordinates and lever coordinates, respectively, and C and S are the cosine and sine, respectively.

Using the transformation matrices, the fixed spring follower positions are transformed into a lever coordinate system and nine circles are generated, as shown in Figure 5b. The PC1 should be designed to have a shape that makes contact with all nine circles at the same time. For the next step, the common circumferential lines and points between adjacent circles are computed to find contact points between the lever and circles. Except for the first and last circles, the ith circle has two common circumferential lines and contact points between the $i - 1^{th}$ and $i + 1^{th}$ circles of spring followers. To assign one tangential line and point to a circle, the line and contact point between the $i - 1^{th}$ circle are selected. As a result, each of the nine circles has a corresponding contact point and slope of the tangential line at that point, as illustrated in Figure 5c.

For the last step, piecewise curves connecting to the adjacent contact points with a satisfactory slope are generated based on the third order Hermite spline. Consequently, the PC1 is formed by connecting the eight piecewise curves that enable tangential contact with the spring follower at each pivot position, thereby achieving subfunction 1 as shown in Figure 5d.

4.2. Partial Contour 2 (PC2): Maximizing the Variable Range of the Compensation Torque

A B-spline curve representation and optimization approach is adopted to generate a PC2 that satisfies the subfunction. As illustrated in Figure 6a, five control points are placed in the lever coordinates system to form a B-spline curve of PC2. The start and end points among the five control points are determined by an obstacle point and a connecting point, respectively, as explained in detail later. A serial link representation is used to avoid the cusp issue [15] as in (6)–(15) as follows:

$$^{L}CP_{0x} = Px_{obstacle} \tag{6}$$

$$^{L}CP_{0y} = Py_{obstacle} \tag{7}$$

$$^{L}CP_{4x} = Px_{connecting} \tag{8}$$

$$^{L}CP_{4y} = Py_{connecting} \tag{9}$$

$$^{L}CP_{1x} = G_1 C_{\Theta_1} + G_2 C_{\Theta_1+\Theta_2} \tag{10}$$

$$^{L}CP_{1y} = G_1 S_{\Theta_1} + G_2 S_{\Theta_1+\Theta_2} \tag{11}$$

$$^{L}CP_{2x} = G_1 C_{\Theta_1} + G_2 C_{\Theta_1+\Theta_2} + G_3 C_{\Theta_1+\Theta_2+\Theta_3} \tag{12}$$

$$^{L}CP_{2y} = G_1 S_{\Theta_1} + G_2 S_{\Theta_1+\Theta_2} + G_3 S_{\Theta_1+\Theta_2+\Theta_3} \tag{13}$$

$$^{L}CP_{3x} = {}^{L}CP_{2x} + G_4 C_{\Theta_1+\Theta_2+\Theta_3+\Theta_4} \tag{14}$$

$$^{L}CP_{3y} = {}^{L}CP_{2y} + G_4 S_{\Theta_1+\Theta_2+\Theta_3+\Theta_4} \tag{15}$$

where G_i is the length of the ith link and Θ_i is the relative angle between the $i - 1^{th}$ and i^{th} link.

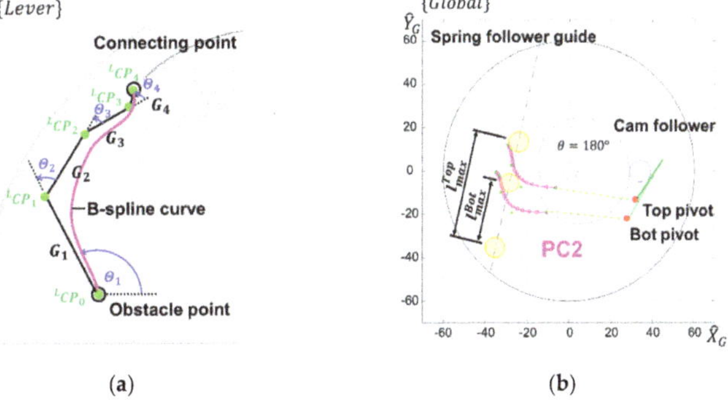

Figure 6. Design of the PC2. The unit of the x- and y-axes are mm: (**a**) B-spline modeling to design PC2 using optimization; (**b**) maximum compression of the spring at each top and bottom pivot position, which constitute the objective function.

The objective function for optimizing the B-spline is defined to maximize the variable range of the compensation torque as follows:

$$g(q) = \frac{l_{max}^{Top}}{l_{max}^{Bot}} \tag{16}$$

l_{max}^{Top} and l_{max}^{Bot} are the maximum deformation of the spring when the pivot is placed at the top and bottom positions, respectively, as shown in Figure 6b. The denominator of the objective function increases as the deformation of the spring at the bottom pivot position decreases, which is related to the lowest compensation torque. The numerator increases when the deformation of the spring at the top pivot position increases, which is related to the highest compensation torque. The case in which the deformation by the PC2 exceeds the allowable deformation of the spring is excluded as a constraint condition. Therefore, the larger the objective function, the larger the variable range of the compensation torque that can be acquired.

Since l_{max}^{Top} and l_{max}^{Bot} can be calculated by the geometrical relation between the B-spline curve and the spring follower guideline, the lever contour defined in the lever coordinate system should be transformed to the global coordinate system. Consequently, the optimization problem is as follows:

$$\begin{aligned} Given\ q &= [G_1, \Theta_1, G_2, \Theta_2, G_3, \Theta_3] \\ maximize\ &g(q) \end{aligned} \tag{17}$$

Table 1 shows the optimization result computed by the genetic algorithm in the global optimization toolbox of MATLAB (MATLAB R2019b, MathWorks, United States). The optimal PC2 generates 50.4 mm and 29.8 mm as the maximum compression lengthx of the spring at the top and bottom pivot positions, respectively.

Table 1. Optimal design variables of PC2.

Pivot	**CP$_1$**	**CP$_2$**	**CP$_3$**
G (m)	0.0162	0.0107	0.0076
Θ (°)	119.5	−62.03	−26.28

4.3. Partial Contour 3 (PC3): Preventing Interference with the Cam Structure

As shown in Figure 4B, the PC3 should be designed to avoid interference with the cam structure only when the cam rotates to 180° at the top pivot position. As illustrated in Figure 7, in the global coordinate system, the common circumferential line and points between the circle of the cam structure and the highest spring follower can be calculated.

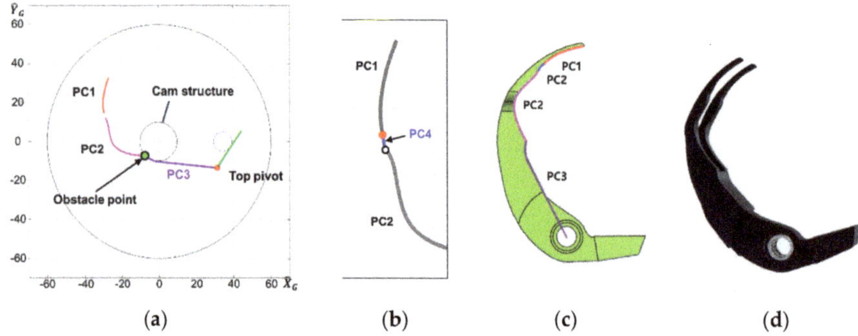

Figure 7. (**a**) Design of PC3 to avoid interference with the cam structure. The units of *x*- and *y*-axes are m; (**b**) design of PC3 to secure C1 continuity between PC1 and PC2; (**c**) final lever contour combining partial contours; (**d**) manufactured lever contour.

The common point on the circle of the cam structure is the obstacle point that is used to design PC2. The obstacle points allow PC2 to avoid interference with the cam structure. Considering that the mass of the lever is proportional to its length, PC3 is determined using the simplest straight line and arc, as illustrated in Figure 7a.

4.4. Partial Contour 4 (PC4): Connecting PC1 and PC2 with C1 Continuity

As a final step of the divide-and-conquer design methodology, combining the derived partial contours with proper continuity condition is important to generate the final contour. The connection between PC2 and PC3 is the area in which the spring follower does not contact with the spring follower. Therefore, the C0 continuity is sufficient to combine the PC2 and PC3. However, the area between the PC1 and PC2 makes contact with the spring follower, so the connection should satisfy the C1 continuity to ensure the smooth motion of the spring follower.

To achieve the connection with C1 continuity, PC4 is designed as follows: On the PC1 side, the starting point and slope at that point are already determined, as described in Section 4.1.

The starting point of PC1 extends by distance ε in the direction of the slope to make the connecting point. The connecting point is used as the fifth control point of PC2, as described in Section 4.2. Using the partial derivatives of PC2, the slope at the connecting point can also be calculated. Finally, based on the Hermite spline representation, PC4 can be determined, as illustrated in Figure 7b. A pseudo code of the algorithm for the design of partial contours is attached in Supplementary Materials.

5. Experiments and Results

The detail design of the lever contour made by combining PC1, PC2, PC3, and PC4 is as presented in Figure 7c. To verify that the lever achieves the desired multifunctions, experiments are separately conducted on each partial contours. As shown in Figure 7d, the lever used in the experiments was manufactured with a 3D-printer (Mark Two, Markforged, MA, US) with carbon-fiber materials that have sufficient strength for experimental verifications. According to the visual measurement method, the experiments successfully checked that each partial contour satisfied the geometrical conditions to achieve a given subfunction. Figure 8 illustrates the manufactured CVGC mechanism with the designed lever contour.

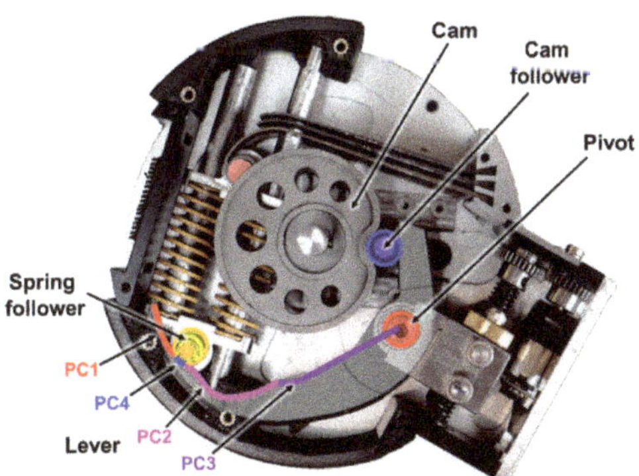

Figure 8. Manufactured CVGC mechanism with the designed lever contour.

To verify the subfunction of PC1, the position of the spring follower was measured by marker tracking when the position of the pivot changed. Although a minor position change of the spring follower occurred, PC1 successfully minimized the movement, as shown in Figure 9.

Figure 9. The visual measurement to verifying PC1 using marker tracking. The red and blue coordinates indicate the position of the spring follower and the pivot position with respect to the origin at the right top of the figure. The circled numbers show the order of pivot variation from the bottom pivot (circled number 1) to the top pivot (circled number 4). The total travel distance of the pivot is 10 mm. Through the proper design of PC1, the position of the spring follower does not be affected by the movement of the pivot.

The position change of the spring follower had 0.4 mm as its maximum value. The reason for the design error of PC1 is the lack of a number of representative pivot points; therefore, this error can be mitigated by increasing the number of representative pivot points and piecewise curve segments of PC1.

Next, each position of the spring follower was measured at the top and bottom pivot positions to verify the function achievement of PC2. As presented in Figure 10 and Table 2, the amounts of spring deformation at each pivot condition are 29.5 and 50.5 mm, which is well-matched with the designed values. The errors of deformation length between simulation and experiment are 1.0% and 0.1% at the bottom (bot) and top pivot conditions, respectively. The absolute errors are 0.7 and 0.1 mm at the bot and top pivot positions, respectively. We expect the position error of marker installation is the main reason for the errors between simulation and experiment.

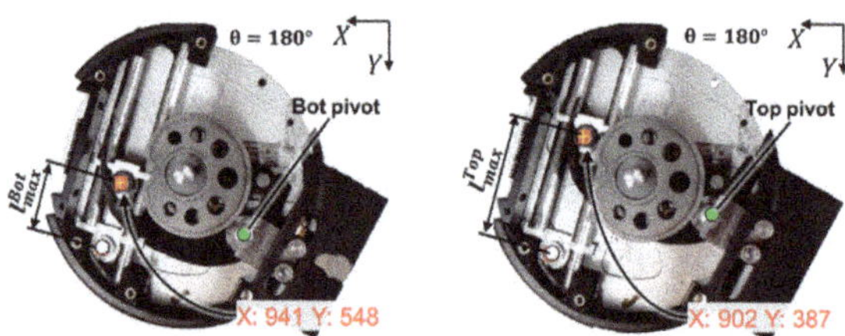

Figure 10. The visual measurement to verifying PC2 using marker tracking. The maximum deformations of the spring at the bot pivot and top pivot conditions are shown in the left and right figures. The red coordinates information indicates the position of the spring follower.

Table 2. Comparison of simulation and experimental results for the maximum deformation of spring at the top and bottom pivot conditions.

Maximum Deformation	Simulation	Experiment
l_{max}^{Bot} (mm)	29.8	29.5
l_{max}^{Top} (mm)	50.4	50.5

Finally, to verify the interference avoidance in the lever and cam structure, which is a function of PC3, the lever was observed at the top pivot position and at 180° cam rotation. As shown on the right side of Figure 10, the lever pushed the spring follower without any interferences. Therefore, PC3 successfully achieves its subfunction.

6. Conclusions

In this study, we propose a contour design methodology for achieving a multifunctional lever based on the morphological principle of the *Salvia pratensis* flower. In this methodology, multiple functions were subdivided into subfunctions. Each subfunction was assigned to a partial contour as a design goal. The final lever contour was determined by combining the partial contours that could achieve the given subfunctions. Through this methodology, a designer can simplify complex and multifunctional curve design problems. Moreover, the most suitable modeling and parameters to determine the method can be selected for each partial contour.

This methodology was specifically explained by a design case study of CVGC, in which the lever should achieve three subfunctions. To achieve the first function, PC1 was modeled using a Hermite spline and was directly calculated to find the design parameters. For the second subfunction, PC2 was determined by B-spline model-based optimization. To satisfy the third subfunction, PC3 was derived simply by using an arc and straight line model. In addition, the PC4 based on Hermite spline was designed to generate a smooth rolling contact with the spring follower.

Finally, through marker tracking-based visual verification, it was verified that, for the derived lever contour, each partial contour could successfully achieve subfunctions. PC1 did not change the position of the spring follower and did not generate deformation of the compression spring. The difference between the maximum deformation of the spring was increased 1.3 times by PC2 when the pivot was placed in the top and bottom (bot) positions as compared with the previously derived lever in [16]. As mentioned in Section 4.2, this difference is directly related to the variable range of compensation torque. PC3 did not prevent any intervention between the lever and cam structure. PC4 achieved C1 continuity between C1 and C2. As future research goals, we plan to install the derived lever into CVGC and measure the performances of the variable gravity compensation torque. In addition, we are developing various versions of CVGC with different sizes and compensation torques. Therefore, we will actively utilize this methodology to design various versions of levers.

Supplementary Materials: The following are available online at https://www.mdpi.com/article/10.3390/app11136015/s1, Figure S1: Pseudo code of the algorithm for the design lever contours for the CVGC case study.

Author Contributions: Conceptualization, J.K., J.M., J.R. and G.L.; methodology, J.K., J.M. and G.L; software, J.K., J.M. and G.L.; validation, J.M., J.R. and G.L.; data analysis, J.K., J.R. and G.L.; writing—original draft preparation, J.K. and G.L.; writing—review and editing, J.K. and G.L.; visualization, J.K. and G.L.; supervision, J.M. and G.L.; project administration, G.L.; funding acquisition, G.L. All authors have read and agreed to the published version of the manuscript.

Funding: This research was funded by the Samsung Research Funding & Incubation Center of Samsung Electronics under Project Number SRFC-IT1903-02 and supported by the Chung-Ang University Research Grants in 2021.

Institutional Review Board Statement: Not applicable.

Informed Consent Statement: Not applicable.

Data Availability Statement: Not applicable.

Conflicts of Interest: The authors declare no conflict of interest.

References

1. Lin, H.; Fang, W. Torsional mirror with an electrostatically driven lever-mechanism. In Proceedings of the 2000 IEEE/LEOS International Conference on Optical MEMS, Kauai, HI, USA, 21–24 August 2000; pp. 113–114.
2. Van der Woude, L.H.V.; Veeger, H.E.J.; de Boer, Y.; Rozendal, R.H. Physiological evaluation of a newly designed lever mechanism for wheelchairs. *J. Med. Eng. Technol.* **1993**, *17*, 232–240. [CrossRef] [PubMed]
3. Li, K.; Gohnert, M. Lever mechanism for vibration isolation. *Appl. Technol. Innov.* **2010**, *1*, 21–28. [CrossRef]
4. Shi, C.; Dong, X.; Yang, Z. A Microgripper with a Large Magnification Ratio and a High Structural Stiffness Based on a Flexure-Enabled Mechanism. *IEEE/ASME Trans. Mechatron.* **2021**. early access. [CrossRef]
5. Kwon, K.; Cho, N.; Jang, W. The design and characterization of a piezo-driven inchworm linear motor with a reduction-lever mechanism. *JSME Int. J. Ser. C* **2004**, *47*, 803–811. [CrossRef]

6. French, M.J.; Widden, M.B. The spring-and-lever balancing mechanism, George Carwardine and the Anglepoise lamp. *Proc. Inst. Mech. Eng. Part C J. Mech. Eng. Sci.* **2000**, *214*, 501–508. [CrossRef]

7. Jafari, A.; Tsagarakis, N.G.; Sardellitti, I.; Caldwell, D.G. A new actuator with adjustable stiffness based on a variable ratio lever mechanism. *IEEE/ASME Trans. Mechatron.* **2012**, *19*, 55–63. [CrossRef]

8. Tsagarakis, N.G.; Sardellitti, I.; Caldwell, D.G. A new variable stiffness actuator (CompAct-VSA): Design and modelling. In Proceedings of the 2011 IEEE/RSJ International Conference on Intelligent Robots and Systems, San Francisco, CA, USA, 25–30 September 2011; pp. 378–383.

9. Kim, B.; Song, J. Hybrid dual actuator unit: A design of a variable stiffness actuator based on an adjustable moment arm mechanism. In Proceedings of the 2010 IEEE International Conference on Robotics and Automation, Anchorage, AK, USA, 3–7 May 2010; pp. 1655–1660.

10. Sun, J.; Guo, Z.; Zhang, Y.; Xiao, X.; Tan, J. A novel design of serial variable stiffness actuator based on an archimedean spiral relocation mechanism. *IEEE/ASME Trans. Mechatron.* **2018**, *23*, 2121–2131. [CrossRef]

11. Groothuis, S.S.; Rusticelli, G.; Zucchelli, A.; Stramigioli, S.; Carloni, R. The Variable Stiffness Actuator vsaUT-II: Mechanical Design, Modeling, and Identification. *IEEE/ASME Trans. Mechatron.* **2014**, *19*, 589–597. [CrossRef]

12. Barrett, E.; Fumagalli, M.; Carloni, R. Elastic Energy Storage in Leaf Springs for a Lever-Arm Based Variable Stiffness Actuator. In Proceedings of the 2016 IEEE/RSJ International Conference on Intelligent Robots and Systems (IROS), Daejeon, Korea, 9–14 October 2016; pp. 537–542.

13. Kim, J.; Moon, J.; Kim, J.; Lee, G. Design of Compact Variable Gravity Compensator (CVGC) Based on Cam and Variable Pivot of a Lever Mechanism. In Proceedings of the 2019 IEEE/RSJ International Conference on Intelligent Robots and Systems (IROS), Macau, China, 3–8 November 2019; pp. 3583–3588.

14. Dežman, M.; Gams, A. Pseudo-linear variable lever variable stiffness actuator: Design and evaluation. In Proceedings of the 2017 IEEE International Conference on Advanced Intelligent Mechatronics (AIM), Munich, Germany, 3–7 July 2017; pp. 785–790.

15. Kim, J.; Moon, J.; Kim, J.; Lee, G. Compact Variable Gravity Compensation Mechanism with a Geometrically Optimized Lever for Maximizing Variable Ratio of Torque Generation. *IEEE/ASME Trans. Mechatron.* **2020**, *25*, 2019–2026. [CrossRef]

16. Kim, J.; Moon, J.; Ryu, J.; Lee, G. CVGC-II: A new version of a compact variable gravity compensator (CVGC) with a wider range of variable torque and an energy-free variable mechanism. *IEEE Trans. Mechatron.* accepted.

17. Reith, M.; Baumann, G.; Claßen-Bockhoff, R.; Speck, T. New insights into the functional morphology of the lever mechanism of *Salvia pratensis* (Lamiaceae). *Ann. Bot.* **2007**, *100*, 393–400. [CrossRef] [PubMed]

applied sciences

Article

Cable Tension Analysis Oriented the Enhanced Stiffness of a 3-DOF Joint Module of a Modular Cable-Driven Human-Like Robotic Arm

Kaisheng Yang [1,2,3,*], Guilin Yang [2,*], Chi Zhang [2], Chinyin Chen [2], Tianjiang Zheng [2], Yuguo Cui [1] and Tehuan Chen [1,4]

[1] School of Mechanical Engineering and Mechanics, Ningbo University, Ningbo 315211, China; cuiyuguo@nbu.edu.cn (Y.C.); chentehuan@nbu.edu.cn (T.C.)

[2] Zhejiang Key Lab of Robotics and Intelligent Manufacturing Equipment Technology, Ningbo Institute of Material Technology and Engineering, Chinese Academy of Sciences (CAS), Ningbo 315201, China; zhangchi@nimte.ac.cn (C.Z.); chenchinyin@nimte.ac.cn (C.C.); zhengtianjiang@nimte.ac.cn (T.Z.)

[3] Zhejiang Marine Development Research Institute, Zhoushan 316021, China

[4] Ningbo Artificial Intelligence Institute, Shanghai Jiao Tong University, Ningbo 315000, China

* Correspondence: yangkaisheng@nbu.edu.cn (K.Y.); glyang@nimte.ac.cn (G.Y.)

Received: 18 November 2020; Accepted: 7 December 2020; Published: 11 December 2020

Abstract: Inspired by the structure of human arms, a modular *cable-driven human-like robotic arm* (CHRA) is developed for safe human–robot interaction. Due to the unilateral driving properties of the cables, the CHRA is redundantly actuated and its stiffness can be adjusted by regulating the cable tensions. Since the trajectory of the *3-DOF joint module* (3DJM) of the CHRA is a curve on Lie group SO(3), an enhanced stiffness model of the 3DJM is established by the covariant derivative of the load to the displacement on SO(3). In this paper, we focus on analyzing the how cable tension distribution problem oriented the enhanced stiffness of the 3DJM of the CHRA for stiffness adjustment. Due to the complexity of the enhanced stiffness model, it is difficult to solve the cable tensions from the desired stiffness analytically. The problem of *stiffness-oriented cable tension distribution* (SCTD) is formulated as a nonlinear optimization model. The optimization model is simplified using the symmetry of the enhanced stiffness model, the rank of the Jacobian matrix and the equilibrium equation of the 3DJM. Since the objective function is too complicated to compute the gradient, a method based on the genetic algorithm is proposed for solving this optimization problem, which only utilizes the objective function values. A comprehensive simulation is carried out to validate the effectiveness of the proposed method.

Keywords: cable-driven robots; human-like robotic arms; human–robot interactions; stiffness adjustment; cable tension analysis

1. Introduction

Unlike conventional robots that work in structured environments, safe human–robot interactions have been a key element for the robots that work in unstructured and unpredictable environments. Inspired by the structure of human arms, a modular *cable-driven human-like robotic arm* (CHRA) is developed for safe human–robot interaction, which employs cables to mimic the functionality of the human muscles. As shown in Figure 1, the CHRA consists of a shoulder joint, an elbow joint and a wrist joint in series, where the shoulder joint and the wrist joint have *three degrees of freedom* (3-DOF) and the elbow joint has *one degree of freedom* (1-DOF). The CHRA employs lightweight cables to drive the rigid links and the cables can be wound into winches mounted onto the base of the CHRA. A *variable-stiffness device* (VSD) is designed and placed along with each driving cable to increase the flexibility of the cables. With these arrangements, the CHRA has flexibility advantages [1], low moving mass [2], large workspace [3] and high payload-to-weight ratio [4]. Due to these advantages, the CHRA is intrinsically safe for human–robot interactions. The proposed CHRA and its joint modules are one kind of *cable-driven mechanisms* (CDMs). In the last decades, various CDMs have been designed for various applications, such as medical robots [5–7], rehabilitation robots [8–12], inspection and repair [13–15], and moving payloads [16–19].

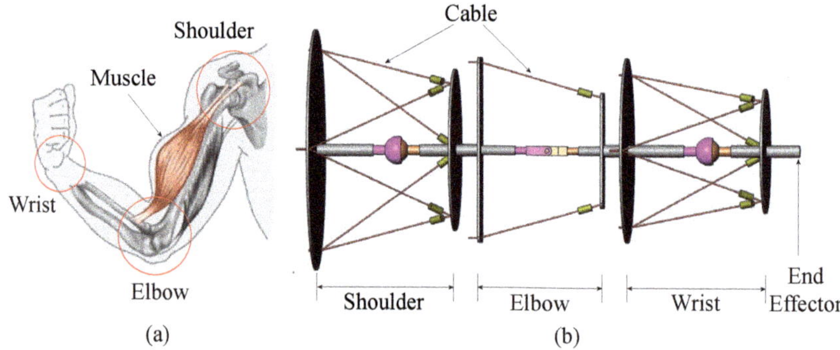

Figure 1. Concept design of the modular cable-driven human-like robotic arm: (**a**) structure of a human arm, (**b**) structure of the *cable-driven human-like robotic arm* (CHRA).

Unlike rigid links, the cables have the unilateral driving property (i.e., they can only pull, but can not push). Hence, the CHRA and its joint modules are redundantly actuated. For a given pose, a lot of cable tension solutions are feasible for the CHRA. Furthermore, the stiffness of the CHRA can be adjusted by regulating the cable tensions. The characteristic of variable stiffness increases the flexibility and safety of the CHRA.

The problem of *stiffness-oriented cable tension distribution* (SCTD) of a CDM aims at finding the optimal cable tensions to achieve a desired stiffness. However, in the last decades, most studies on cable tension distribution have been carried out to minimize the cable tensions to reduce the energy consumption [20–27]. In [28], the SCTD problem of a CDM was studied by formulating it as an optimization model, and a gradient-projection-based algorithm was presented to solve the optimization problem. However, it utilized the determinant of the stiffness matrix as the objective function. With this method, the desired stiffness cannot be achieved accurately. In [29], the SCTD problem of a 3-DOF cable-driven spherical mechanism was studied and it established the optimization model with all entries of the stiffness matrix, not only the determinant of the stiffness matrix. However, it employed the conventional stiffness model of the CDM, which was derived by the conventional differential formula of the load to the displacement. Since the trajectory of the 3-DOF cable-driven spherical mechanism is a curve on Lie group SO(3) and SO(3) is nonlinear, the stiffness model based on the conventional differential formula is not exactly accurate on SO(3). Its stiffness should be evaluated by the variation of its load against its displacement on SO(3).

In order to study the SCTD problem of the proposed CHRA, the SCTD problems of the 1-DOF and 3-DOF joint modules should be studied first. Due to the simple structure of the 1-DOF joint module, its SCTD problem can be solved easily. For the *3-DOF joint module* (3DJM) of the CHRA, its trajectory is a curve on SO(3). We derive the stiffness model of the 3DJM by the covariant derivative of the load to the displacement on SO(3), and call it the enhanced stiffness model of the 3DJM. In this paper, we focus on investigating how the cable tension distribution problem oriented the enhanced stiffness of the 3DJM. Due the complexity of the enhanced stiffness model, it is difficult to solve the cable tensions from the desired stiffness of the 3DJM analytically. We formulate the SCTD problem of the 3DJM as a nonlinear constrained optimization problem. Due to the symmetry of the enhanced stiffness model of the 3DJM, the desired stiffness matrix can be transformed to a diagonal matrix by an orthogonal transformation. Based on the analysis of the rank of the Jacobian matrix of the 3DJM, the six cables can be divided into two groups: one group with three cables for position adjustment by regulating the cable lengths, and another group with the remaining three cables for stiffness adjustment by regulating the cable tensions. In this manner, the position and stiffness of the 3DJM can be adjusted simultaneously. Furthermore, three cable tensions for position adjustment can be expressed by another three cable tensions for stiffness adjustment using the equilibrium equation of the 3DJM. That means the six decision variables of the optimization model can be reduced to three. Since the objective function of the optimization model is too complicated to compute the gradient, a direct optimization method based on the genetic algorithm is proposed for solving the optimization problem, which only utilizes the objective function values. A comprehensive simulation is carried out to validate the effectiveness of the proposed method. The results show that the proposed method provides an accurate and efficient way to adjust the stiffness of the 3DJM by regulating the cable tensions. In summary, the main contribution of this paper is solving the cable tension distribution problem with the enhanced stiffness model of the 3DJM using a method based on the genetic algorithm.

2. Enhanced Stiffness Model of the 3DJM

As shown in Figure 2, the 3-DOF joint module (3DJM) of the CHRA was made up with a base, a moving-platform and a passive spherical joint connecting them. Six cables were employed to drive the moving-platform and a *variable-stiffness device* (VSD) was placed along with each driving cable. For routing the cables, six small holes were drilled in the moving-platform and the base, denoted by A_i and B_i ($i = 1, 2, \ldots, 6$), respectively. In this design, $A_2 A_3 = A_4 A_5 = A_6 A_1 = l_A$, $B_1 B_2 = B_3 B_4 = B_5 B_6 = l_B$, $A_1 A_2 = A_3 A_4 = A_5 A_6 = e_A$, $B_2 B_3 = B_4 B_5 = B_6 B_1 = e_B$, where $e_A \approx 0$ and $e_B \approx 0$. The distance between the moving-platform and the spherical joint is denoted as $h_A \in \mathbb{R}$, and the distance between the base and the spherical joint is denoted as $h_B \in \mathbb{R}$. Each cable was actuated by a cable driving unit.

To describe the pose of the 3DJM, two frames were attached to the moving-platform (named moving frame {A}) and the base (named base frame {B}), which were both located at the center of the joint O. With the two frames, the pose of the 3DJM can be represented by a rotation matrix $R \in SO(3)$. Thus, the motion trajectory of the 3DJM was a parameterized curve $R(t)$ on $SO(3)$. The enhanced stiffness model of the 3DJM was established by using the covariant derivative of the load to the displacement on SO(3) [30].

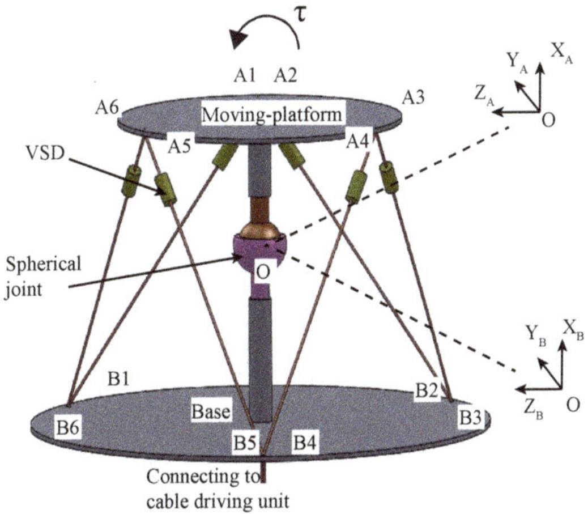

Figure 2. Concept design of the *3-DOF joint module* (3DJM).

2.1. Displacement and Load of the 3DJM on SO(3)

According to the exponential formula [31], the trajectory curve of the 3DJM $R(t)$ yields

$$R(t) = e^{\hat{\zeta}(t)}, \tag{1}$$

where $\hat{\zeta} \in so(3)$ is an element of Lie algebra $so(3)$ and satisfies

$$\zeta = \begin{pmatrix} \zeta_1 \\ \zeta_2 \\ \zeta_3 \end{pmatrix} \rightarrow \hat{\zeta} = \begin{pmatrix} 0 & -\zeta_3 & \zeta_2 \\ \zeta_3 & 0 & -\zeta_1 \\ -\zeta_2 & \zeta_1 & 0 \end{pmatrix}. \tag{2}$$

Here, $\zeta \in \mathbb{R}^3$ is called the coordinates of $\hat{\zeta} \in so(3)$. The derivative of $R(t)$ to time t, i.e., $\dot{R}(t)$, is an element of the tangent space of $SO(3)$ at the point $R(t)$, denoted as $T_{R(t)}SO(3)$. According to (1), $\dot{R}(t)$ satisfies the following equation

$$\dot{R}(t) = R(t)\hat{\dot{\zeta}}(t) = R(t)\hat{\omega}(t), \tag{3}$$

where $\omega(t) = \dot{\zeta}(t) \in so(3)$ is the angular velocity of the 3DJM. Given $\hat{e}_j$ $(j = 1, 2, 3)$ as the basis of $so(3)$, where $e_1 = (1, 0, 0)^T$, $e_2 = (0, 1, 0)^T$ and $e_3 = (0, 0, 1)^T$, then $\dot{R}(t)$ can be expressed as

$$\dot{R}(t) = R(t)\hat{\omega}(t) = R(t)\sum_{j=1}^{3}\omega^j\hat{e}_j = \sum_{j=1}^{3}\omega^j R(t)\hat{e}_j = \sum_{j=1}^{3}\omega^j L_j|_{R(t)}, \tag{4}$$

where $L_j|_{R(t)} = R(t)\hat{e}_j$ $(j = 1, 2, 3)$ are taken as the basis of $T_{R(t)}SO(3)$ and $L_j|_{R(t)}$ are always written as L_j for short. From (4), it can be concluded that $T_{R(t)}SO(3)$ is isomorphic to $so(3)$.

Since $\dot{R}(t)$ describes the velocity of the 3DJM, the instantaneous displacement of the 3DJM can be represented by δR, yielding

$$\delta R(t) = \dot{R}(t)\delta t. \tag{5}$$

That means the instantaneous displacement of the 3DJM can be described on the tangent space of $SO(3)$. Similarly, since the moment load $\hat{\tau}$ is an element of $so^*(3)$, the dual space of $so(3)$, by analogy with $\hat{\omega} \in so(3)$ and $\dot{R}(t) \in T_{R(t)}SO(3)$, we define $F(t) = R(t)\hat{\tau}(t)$ as an element of the dual space of the tangent space $T_{R(t)}SO(3)$ at $R(t)$, called *cotangent space* and denoted as $T^*_{R(t)}SO(3)$. The element of the cotangent space is named as the cotangent vector. Given $\hat{\lambda}^k$ $(k = 1, 2, 3)$ as the basis of $so^*(3)$ and $\Lambda^k|_{R(t)} = R(t)\hat{\lambda}^k$ $(k = 1, 2, 3)$ as the basis of $T^*_{R(t)}SO(3)$, then $F(t) \in T^*_{R(t)}SO(3)$ satisfies

$$F(t) = R(t)\hat{\tau}(t) = \sum_{k=1}^{3}\tau_k R(t)\hat{\lambda}^k = \sum_{k=1}^{3}\tau_k \Lambda^k|_{R(t)}. \tag{6}$$

It shows that $T^*_{R(t)}SO(3)$ is isomorphic to $so^*(3)$. Thus, $F(t)$ can be employed to represent the load applied on the 3DJM. That means the load applied on the 3DJM can be described on the cotangent space of $SO(3)$.

2.2. Enhanced Stiffness Model of the 3DJM on SO(3)

According to the definition, the stiffness of the 3DJM is evaluated by the variation of its load against its displacement. However, the loads and displacements for different poses of the 3DJM are in different cotangent spaces and tangent spaces of SO(3). Considering the property of the motion of the 3DJM, an affine connection called *Levi-Civita* connection is introduced into $SO(3)$ to connect different tangent spaces on SO(3) [32]. Given a curve $R(t)$ and a vector field $V(t)$ on $SO(3)$, the affine connection specifies how a vector $V(t_0)$ in the tangent space at point $R(t_0)$ can be mapped to another vector $V^{t_0}(t_1)$ of the tangent space at some other point $R(t_1)$. The vector $V^{t_0}(t_1)$ is called the parallel transport of $V(t_0)$ along the curve $R(t)$ [33]. Then, a differentiation operation can be defined on SO(3) as below

$$\nabla_{\dot{R}} V\big|_{R(t_0)} = \lim_{t_1 \to t_0} \frac{V^{t_0}(t_1) - V(t_0)}{t_1 - t_0}, \tag{7}$$

where $\nabla_{\dot{R}} V\big|_{R(t_0)}$ is called the *covariant derivative* of $V(t)$ along the curve $R(t)$. For two vectors, V_1 and V_2, in the tangent spaces of SO(3), $\nabla_{V_2} V_1$ is employed to represent the covariant derivative of V_1 in the direction V_2. For a real-valued function f on $SO(3)$, the covariant derivative $\nabla_V f$ is usually written as $V \circ f$. According to the property of covariant derivative, the covariant derivative of the load $F(t)$ to the instantaneous displacement $\delta R(t)$ can be written as

$$\nabla_{\delta R(t)} F = \nabla_{\dot{R}(t)\delta t} F = \delta t \nabla_{\sum_{j=1}^3 \omega^j L_j} F = \delta t \sum_{j=1}^{3} \omega^j \nabla_{L_j} F = \delta t K \omega = K \delta \zeta, \tag{8}$$

and $K = (\nabla_{L_1} F, \nabla_{L_2} F, \nabla_{L_3} F) \in \mathbb{R}^{3 \times 3}$ is employed as the stiffness of the 3DJM. The components of the stiffness matrix K, i.e., K_{jk} $(j, k = 1, 2, 3)$, yield

$$K_{jk} = \langle \nabla_{L_k} F, L_j \rangle. \tag{9}$$

According to the property of the Levi–Civita connection [30], we have

$$K_{jk} = L_k \circ \langle F, L_j \rangle - \langle F, \nabla_{L_k} L_j \rangle = L_k \circ \tau_j - \frac{1}{2} \sum_{r=1}^{3} \tau_r \gamma_{kj}^r, \tag{10}$$

where the coefficients γ_{kj}^r $(j, k, r = 1, 2, 3) \in \mathbb{R}$ are zero except

$$\gamma_{12}^3 = \gamma_{31}^2 = \gamma_{23}^1 = 2, \quad \gamma_{21}^3 = \gamma_{13}^2 = \gamma_{32}^1 = -2. \tag{11}$$

Furthermore, since the 3DJM is a conservative mechanical system, its stiffness matrix K can be proved to be symmetric at every pose, even if there is an external load applied on it [30].

2.3. Parametric Stiffness Formulation of the 3DJM

In Figure 2, $a_i = \overrightarrow{OA_i}, b_i = \overrightarrow{OB_i} \in \mathbb{R}^3$ $(i = 1, 2, \ldots, 6)$ are the position vectors of points A_i and B_i, respectively. The vector from A_i to B_i along the ith cable, denoted as $c_i = \overrightarrow{A_i B_i}$, yields

$$c_i = b_i - a_i = c_i u_i \in \mathbb{R}^3, \tag{12}$$

where $c_i = \|c_i\| \in \mathbb{R}$ is the length of the cable from A_i to B_i and $u_i = c_i / c_i \in \mathbb{R}^3$ is the unitary vector of c_i.

Let $t_i = t_i u_i \in \mathbb{R}^3$ $(i = 1, 2, \ldots, 6)$ be the cable tension vector of the ith cable, where $t_i = \|t_i\| \in \mathbb{R}^3$ is the value of the cable tension. According to the equilibrium equation, the moment load τ applied on the moving platform with respect to the point O satisfies

$$\tau + \sum_{i=1}^{6} a_i \times t_i = 0. \tag{13}$$

Denoting $T = (t_1, t_2, \ldots, t_6)^T \in \mathbb{R}^6$ as the vector of six cable tension values, the moment load τ can be written as

$$\tau = -\sum_{i=1}^{6} a_i \times t_i = -\sum_{i=1}^{6} (a_i \times u_i) t_i = J^T T, \tag{14}$$

where $J \in \mathbb{R}^{6 \times 3}$ is called the Jacobian matrix of the 3DJM and yields

$$J = -(a_1 \times u_1, a_2 \times u_2, \cdots, a_6 \times u_6)^T. \tag{15}$$

Denoting $S = J^T \in \mathbb{R}^{3 \times 6}$ and S_{ji} $(j = 1,2,3; i = 1,2,\ldots,6)$ as the components of S, the components of τ yield

$$\tau_j = \sum_{i=1}^{6} S_{ji} t_i. \tag{16}$$

Then, the expression $L_k \circ \tau_j$ in (10) can be written as

$$L_k \circ \tau_j = \lim_{\Delta t \to 0} \frac{\tau_j\big|_{R(t+\Delta t)} - \tau_j\big|_{R(t)}}{\omega^k \Delta t} = \frac{\partial \tau_j}{\partial \zeta^k}\bigg|_{R(t)} = \sum_{i=1}^{6} \left(\frac{\partial S_{ji}}{\partial \zeta^k} t_i + S_{ji} \frac{\partial t_i}{\partial \zeta^k} \right). \tag{17}$$

where $\tau_j\big|_{R(t)}$ $(j = 1,2,3)$ are the components of τ when the 3DJM stays at the pose $R(t)$.

Writing the six cable lengths as a vector $c = (c_1, c_2, \ldots, c_6)^T \in \mathbb{R}^6$, then we have the following equation according to the principle of virtual work, i.e.,

$$T^T \delta c = \tau^T \delta \zeta. \tag{18}$$

Substituting (14) into (18), we have

$$\delta c = J \delta \zeta. \tag{19}$$

It shows that $\frac{\partial c_i}{\partial \zeta^k} = J_{ik}$ $(i = 1,2,\ldots,6; k = 1,2,3)$, where $J_{ik} \in \mathbb{R}$ is the component of $J \in \mathbb{R}^{6 \times 3}$. Writing $k_i = \frac{\partial t_i}{\partial c_i}$ as the stiffness of the ith cable, the expression $\frac{\partial t_i}{\partial \zeta^k}$ in (17) satisfies

$$\frac{\partial t_i}{\partial \zeta^k} = \frac{\partial t_i}{\partial c_i} \frac{\partial c_i}{\partial \zeta^k} = k_i J_{ik}. \tag{20}$$

Substituting (16), (17) and (20) into the stiffness model (10), K_{jk} can be written as

$$K_{jk} = \sum_{i=1}^{6} \left(\frac{\partial S_{ji}}{\partial \zeta^k} t_i + k_i S_{ji} J_{ik} \right) - \frac{1}{2} \sum_{r=1}^{3} \sum_{i=1}^{6} \gamma_{kj}^r S_{ri} t_i. \tag{21}$$

Since $S = J^T$, the parametric formulation of the enhanced stiffness model (10) is given below

$$K = D + S K_{\text{diag}} S^T - \widehat{ST}. \tag{22}$$

where $D = \left(\frac{\partial S}{\partial \zeta^1} T, \frac{\partial S}{\partial \zeta^2} T, \frac{\partial S}{\partial \zeta^3} T \right)$ and $K_{\text{diag}} = \text{diag}\{k_1, k_2, \cdots, k_6\}$.

3. Modeling of the Cable Tension Distribution Problem Oriented the Enhanced Stiffness of the 3DJM

For a given pose, the Jacobian matrix of the 3DJM is constant. The stiffness of the 3DJM only depends on the cable tensions according to (22). That means the stiffness of the 3DJM can be adjusted by regulating the cable tensions. Given a desired feasible stiffness matrix $K_{\text{des}} \in \mathbb{R}^{3 \times 3}$ for the 3DJM at a given pose R with a load τ, the corresponding cable tensions should be solved from the following equation

$$D + S K_{\text{diag}} S^T - \widehat{ST} = K_{\text{des}}. \tag{23}$$

Due to the complexity of the stiffness formulation of the 3DJM, it is difficult to solve the cable tensions $T \in \mathbb{R}^6$ from (23) analytically.

As the stiffness K_{des} is a 3×3 real symmetric matrix in the frame {A}, there exists a real orthogonal matrix $Q \in \mathbb{R}^{3 \times 3}$ to transfer the stiffness matrix from the frame {A} to another frame {E}, in which the symmetric stiffness matrix can be represented by a diagonal matrix [34], i.e.,

$$Q^T K_{des} Q = {}^E K_{des} = \mathbf{diag}\{K_{d1}, K_{d2}, K_{d3}\}, \tag{24}$$

where the orthogonal matrix $Q \in SO(3)$ satisfies

$$Q^T Q = Q Q^T = I_{3 \times 3}. \tag{25}$$

and ${}^E K_{des}$ represents the desired stiffness in the frame {E}. Here, K_{di} $(i = 1, 2, 3)$ is the diagonal component of the diagonal matrix ${}^E K_{des}$.

Since the desired stiffness is a diagonal matrix in the frame {E}, we can discuss the SCTD problem in the frame {E} to make it simple. On the other hand, as the actual stiffness matrix K_{act} may not be exactly symmetric, a symmetric stiffness matrix $\overline{K}_{act}$ can be obtained as below

$$\overline{K}_{act} = \frac{1}{2}(K_{act} + K_{act}^T). \tag{26}$$

In the frame {E}, the symmetric actual stiffness $\overline{K}_{act}$ is represented by ${}^E \overline{K}_{act}$, which yields

$${}^E \overline{K}_{act} = Q^T \overline{K}_{act} Q = Q^T (D + S K_{diag} S^T - \widehat{ST}) Q. \tag{27}$$

Denote K_{ai} $(i = 1, 2, 3)$ as the diagonal elements of ${}^E \overline{K}_{act}$, the SCTD problem can be formulated as an optimization model

$$
\begin{aligned}
\text{Minimize} \quad & g(T) = \sqrt{\sum_{i=1}^{3} (K_{ai}(T) - K_{di})^2}, \\
\text{Subject to} \quad & ST = \tau, \\
& T_{min} \preceq T \preceq T_{max},
\end{aligned}
\tag{28}
$$

where $T_{min} \in \mathbb{R}^6$ represents the minimum of the tension vector T to avoid the cable be slack and $T_{max} \in \mathbb{R}^6$ represents the maximum of the tension vector T to avoid the cable tensions exceeding the capability of the cable driving units. Here, $X \preceq Y$ $(X, Y \in \mathbb{R}^n)$ represents that each element of X is no more than the corresponding element of Y.

According to analysis of the rank of the Jacobian matrix of the 3DJM in the Appendix A, we have $\mathbf{rank}(S) = 3$. Then, the column vectors of $S \in \mathbb{R}^{3 \times 6}$ can be divided into two parts, $S_p, S_s \in \mathbb{R}^{3 \times 3}$. Here S_p yields $\mathbf{rank}(S_p) = 3$ and it is called a basis of matrix S. Correspondingly, the cable tension vector $T \in \mathbb{R}^6$ can be divided into two parts, $T_p, T_s \in \mathbb{R}^3$. That means the six driving cables can be divided into two groups: one group with three cables for position adjustment by regulating the cable lengths, and another group with the remaining three cables for stiffness adjustment by regulating the cable tensions. In this manner, the position and stiffness of 3DJM can be adjusted simultaneously. According to equilibrium Equation (14), the tensions for the position adjustment cables, i.e., T_p, can be represented by the tensions for the stiffness adjustment cables, i.e., T_s, as follows

$$T_p = S_p^{-1} \tau - S_p^{-1} S_s T_s. \tag{29}$$

Substituting T_p into the optimization model (28), the objective function $g(T)$ is simplified to $h(T_s)$, i.e.,

$$g(T) = g(T_p, T_s) = h(T_s), \tag{30}$$

and the linear inequality constraints are written as

$$\Phi T_s \preceq w, \tag{31}$$

where

$$\Phi = \begin{pmatrix} I_{3\times3} \\ -I_{3\times3} \\ -S_p^{-1}S_s \\ S_p^{-1}S_s \end{pmatrix}, \tag{32}$$

and

$$w = \begin{pmatrix} T_{s-max} \\ -T_{s-min} \\ -S_p^{-1}\tau + T_{p-max} \\ S_p^{-1}\tau - T_{p-min} \end{pmatrix}. \tag{33}$$

Here, $T_{p-min}, T_{s-min} \in \mathbb{R}^3$ represent the minimum of the tension vector T_p, T_s, respectively, and $T_{p-max}, T_{s-max} \in \mathbb{R}^3$ represent the maximum of the tension vector T_p, T_s, respectively. Then, the optimization model (28) is simplified to an equivalent model, where six decision variables are reduced to three and the linear equality constraint are eliminated, i.e.,

$$\text{Minimize} \quad h(T_s) = \sqrt{\sum_{i=1}^{3}(K_{ai}(T_s) - K_{di})^2}, \tag{34}$$

$$\text{Subject to} \quad \Phi T_s - w \preceq 0.$$

4. Cable Tension Solution Based on the Genetic Algorithm for the 3DJM

Since the objective function of the optimization model (34) is complicated and the stiffness of the VSD relative to the cable tension does not need to be different at each point, the widely used gradient-based algorithms for nonlinear optimization problems are not suitable for this optimization model. The direct optimization methods, which do not need the gradient of the objective function, can be employed for this optimization model, such as Complex method, Nelder–Mead algorithm [35] and genetic algorithm. In this paper, a generic method based on the genetic algorithm is proposed to solve the nonlinear constrained optimization model.

A genetic algorithm is inspired by biological evolutionary theory. It is an iterative procedure which usually operates on a population of constant size [36]. In order to apply the genetic algorithm, we revise the optimization model (34) as follows

$$\text{Maximize} \quad f(T_s) = \frac{1}{h(T_s)}, \tag{35}$$

$$\text{Subject to} \quad \Phi T_s - w \preceq 0.$$

where T_s is taken as an individual (also called a *chromosomes*), and $f(T_s)$ is taken as the fitness function of the individual T_s.

The genetic algorithm is a stochastic iterative algorithm, where each iteration step is also called a *generation*. Since the genetic algorithm cannot guarantee convergence [36], the termination condition of the proposed method is commonly triggered by finding an acceptable solution for the problem or by reaching a maximum number of generations. Here, we define a parameter $\eta \in \mathbb{R}$ to evaluate the closeness of two matrices $K_1, K_2 \in \mathbb{R}^{3\times3}$,

$$\eta(K_1, K_2) = \frac{v_1 \cdot v_2}{\|v_1\|\|v_2\|} \times 100\%, \tag{36}$$

where v_i $(i = 1, 2)$ is the vector form of K_i and $\|v_i\|$ is the norm of v_i. The iterative procedure of the proposed method is shown below, and it terminates when the following condition achieves

$$1 - \eta(\overline{K}_{\text{act}}, K_{\text{des}}) \leq 0.0005. \tag{37}$$

Step 1: *Generate an initial population*

Generate an initial population of individuals randomly or heuristically. Each individual can be represented by a binary string.

Step 2: *Evaluate all individuals*

Compute T_p via (29), and evaluate the fitness function $f(T_s)$ for the individuals of the current population. If T_p does not satisfy $T_{p-min} \preceq T_p \preceq T_{p-max}$, set the fitness value as zero.

Step3: *Check termination condition*

Check if the current population satisfies the termination condition. Stop the iterative procedure if it satisfies, and generate a new population if not.

Step 4: *Selective reproduction*

Select the individuals of the current population (usually with a probability proportional to their relative fitness values) and produce offspring candidates.

Step 5: *Crossover and mutation*

Perform two operators, named crossover and mutation, on the above offspring candidates for producing a new population. Execute **Step 3** for the new population.

- *Crossover* is the primary genetic operator, which swaps the substrings of two individuals, called parents, before and after a randomly selected crossover point to produce two new individuals, called offsprings.
- *Mutation* is essentially an arbitrary modification, which flips bits randomly in a string with a certain probability called the mutation rate.

The diagram of the proposed method is shown in Figure 3.

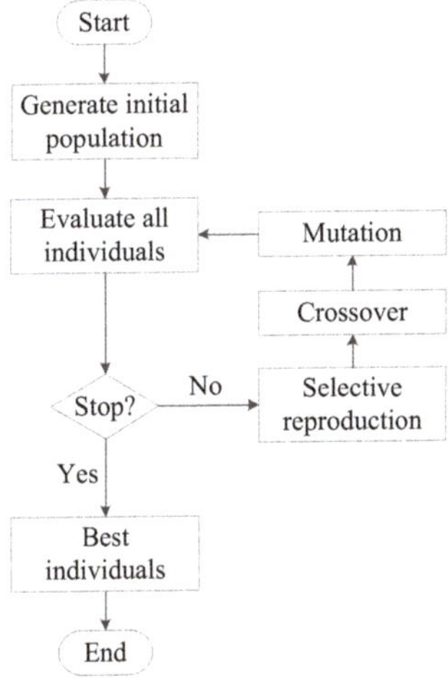

Figure 3. Diagram of the proposed method for the stiffness-oriented cable tension distribution problem of the 3DJM.

5. Simulation Examples

In order to validate the proposed method, a simulation was carried out on a certain 3DJM. As shown in Figure 4, the dimensions of the 3DJM are given by $l_A = 0.10$ [m], $l_B = 0.13$ [m], $e_A = e_B = 0.002$ [m] and $h_A = h_B = 0.08$ [m].

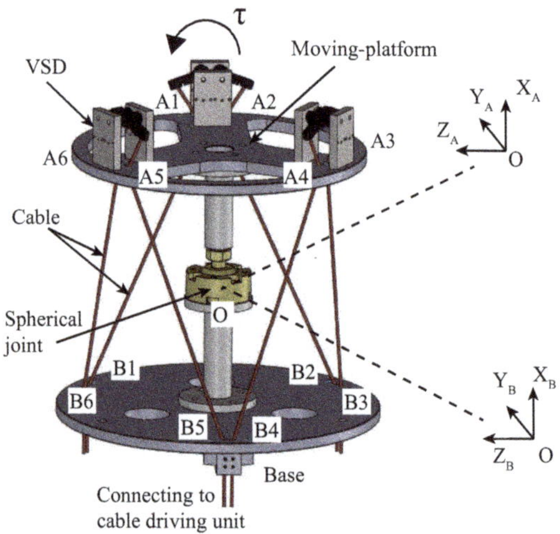

Figure 4. CAD model of the 3DJM.

The VSDs were designed to fix onto the moving platform of the 3DJM to extend the range of stiffness variation of each cable. The CAD model of the VSD is shown in Figure 5a. According to the diagram of the VSD, as shown in Figure 5b, the length of the cable in the VSD yields

$$l = \sqrt{h^2 + d^2 - 2hd \cos\phi}, \tag{38}$$

and the cable tension t_c applied on the VSD yields

$$t_c = \frac{l k_s (\phi_0 - \phi)}{hd \sin\phi}, \tag{39}$$

where $h \in \mathbb{R}$ is the height of the revolute joint, $d \in \mathbb{R}$ is the length of the rigid-link, $k_s \in \mathbb{R}$ is the stiffness of the spring, $\phi \in \mathbb{R}$ is the angle of the rigid-link, and $\phi_0 \in \mathbb{R}$ is the initial value of ϕ. Then the stiffness of the VSD, denoted as $k_{VSD} \in \mathbb{R}$, can be described by

$$k_{VSD} = \frac{dt_c}{dl} = \frac{dt_c}{d\phi} \bigg/ \frac{dl}{d\phi}. \tag{40}$$

Given the parameters of the designed VSD, i.e., $k_s = 1.20$ Nm/rad, $d = 0.018$ m, $h = 0.03$ m, and $\phi_0 = 0.53$ rad, the stiffness of the VSD can be approximated by a polynomial expression

$$k_{VSD} \approx 8.005 t_c^2 - 239.4 t_c + 5415. \tag{41}$$

The corresponding stiffness–tension curve is shown in Figure 5c. For this 3DJM, the lower limits of the cable tensions are given by

$$T_{min} = (10, 10, 10, 10, 10, 10)^T \ [N],$$

and the upper limits of the cable tensions are given by

$$T_{\max} = (400, 400, 400, 400, 400, 400)^T \; [N].$$

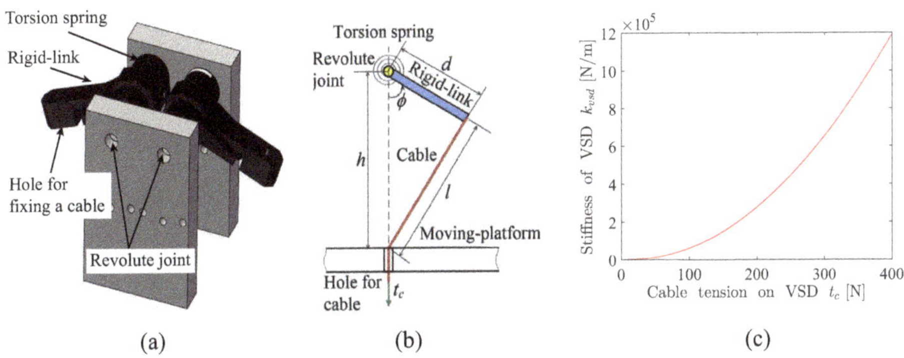

Figure 5. *variable-stiffness device* (VSD) for simulation: (**a**) CAD model of the VSD, (**b**) diagram of the VSD, (**c**) stiffness–tension curve of the VSD.

The simulation examples are implemented for two different desired poses R_{des1} and R_{des2} with different loads, where

$$R_{des1} = \begin{pmatrix} 0.992 & -0.060 & -0.111 \\ 0.090 & 0.952 & 0.292 \\ 0.088 & -0.300 & 0.950 \end{pmatrix}, \quad R_{des2} = \begin{pmatrix} 0.950 & 0.088 & -0.300 \\ -0.111 & 0.992 & -0.060 \\ 0.292 & 0.090 & 0.952 \end{pmatrix}. \tag{42}$$

For each desired pose, three desired stiffness matrices are given. Thus, the simulation examples can be divided into six sub-cases, as listed in Table 1.

Table 1. Six sub-cases of the simulation with different desired poses, applied loads and desired stiffness matrices.

Case	Pose R_{des}	Load τ [Nm]	Desired Stiffness K_{des} [Nm/rad]
Case 1a	R_{des1}	$\tau_{1a} = \begin{pmatrix} 11.62 \\ 8.09 \\ 4.05 \end{pmatrix}$	$K_{des1a} = \begin{pmatrix} 1806.79 & 323.63 & 765.79 \\ 323.63 & 443.92 & -49.61 \\ 765.79 & -49.61 & 634.78 \end{pmatrix}$
Case 1b	R_{des1}	$\tau_{1b} = \begin{pmatrix} 9.26 \\ 11.39 \\ 4.55 \end{pmatrix}$	$K_{des1b} = \begin{pmatrix} 1131.24 & 160.46 & 62.27 \\ 160.46 & 1918.78 & -134.28 \\ 62.27 & -134.28 & 588.78 \end{pmatrix}$
Case 1c	R_{des1}	$\tau_{1c} = \begin{pmatrix} 7.24 \\ 6.60 \\ 2.72 \end{pmatrix}$	$K_{des1c} = \begin{pmatrix} 1600.13 & -353.41 & 320.28 \\ -353.41 & 1687.70 & 244.90 \\ 320.28 & 244.90 & 477.74 \end{pmatrix}$
Case 2a	R_{des2}	$\tau_{2a} = \begin{pmatrix} 2.95 \\ 8.05 \\ 2.08 \end{pmatrix}$	$K_{des2a} = \begin{pmatrix} 1302.42 & -140.27 & 518.72 \\ -140.27 & 617.78 & 147.18 \\ 518.72 & 147.18 & 503.89 \end{pmatrix}$
Case 2b	R_{des2}	$\tau_{2b} = \begin{pmatrix} -7.64 \\ 9.56 \\ -5.90 \end{pmatrix}$	$K_{des2b} = \begin{pmatrix} 1670.44 & 180.42 & 499.26 \\ 180.42 & 1226.99 & -350.38 \\ 499.26 & -350.38 & 676.84 \end{pmatrix}$
Case 2c	R_{des2}	$\tau_{2c} = \begin{pmatrix} 3.70 \\ 14.72 \\ -2.58 \end{pmatrix}$	$K_{des2c} = \begin{pmatrix} 1563.55 & 298.57 & 15.13 \\ 298.57 & 2395.40 & -280.72 \\ 15.13 & -280.72 & 687.41 \end{pmatrix}$

Here, we take Case 1a as an example to perform the proposed method. The other five sub-cases are similar to Case 1a. In Case 1a, the rotation matrix Q is computed according to (24),

$$Q = \begin{pmatrix} 0.159 & -0.432 & 0.888 \\ -0.727 & 0.557 & 0.401 \\ 0.668 & 0.710 & 0.225 \end{pmatrix}, \tag{43}$$

which transforms K_{des1a} to a diagonal matrix $^{E}K_{\text{des1a}} = \mathbf{diag}\{-975, -2937, -2756\}$ [Nm/rad]. According to (15), the transpose of the Jacobian matrix J_1 of the 3DJM at the given pose R_{des1} is computed

$$S_1 = J_1^T = \begin{pmatrix} 0.05 & -0.02 & -0.04 & -0.01 & 0.03 & 0.06 \\ -0.04 & -0.04 & 0 & 0.05 & 0.05 & 0.01 \\ -0.01 & 0.02 & -0.02 & 0.01 & -0.03 & 0.03 \end{pmatrix}. \tag{44}$$

The matrix S_1 is divided into two parts as below

$$S_{p1} = \begin{pmatrix} 0.05 & -0.04 & 0.03 \\ -0.04 & 0 & 0.05 \\ -0.01 & -0.02 & -0.03 \end{pmatrix}, \quad S_{s1} = \begin{pmatrix} -0.02 & -0.01 & 0.06 \\ -0.04 & 0.05 & 0.01 \\ 0.02 & 0.01 & 0.03 \end{pmatrix}. \tag{45}$$

The proposed method is implemented to solve the optimization model (34) to find out the cable tension distribution for the desired stiffness K_{des1a}. The iteration curve of the proposed method for Case 1a is shown in Figure 6a, which shows that the optimization process achieves the termination condition within 7 generations and the optimal cable tension distribution for the desired stiffness K_{des1a} is

$$T_{\text{opt1a}} = (20.84, 72.86, 139.73, 83.88, 123.42, 248.62)^T \ [N].$$

The corresponding actual stiffness K_{act1a} and $\overline{K}_{\text{act1a}}$ are computed as below

$$K_{\text{act1a}} = \begin{pmatrix} 1842.37 & 339.16 & 787.59 \\ 324.04 & 388.59 & -29.52 \\ 810.48 & -61.30 & 640.20 \end{pmatrix}, \tag{46}$$

$$\overline{K}_{\text{act1a}} = \begin{pmatrix} 1842.37 & 331.60 & 799.03 \\ 331.60 & 388.59 & -45.41 \\ 799.03 & -45.41 & 640.20 \end{pmatrix}. \tag{47}$$

According to (36), in Case 1a, $\eta(\overline{K}_{\text{act1a}}, K_{\text{des1a}}) = 99.9554\%$. It shows that, when the cable tensions achieve T_{opt1a}, the actual stiffness approaches the desired stiffness K_{des1a}.

The results of the simulation for all the six sub-cases are summarized in Table 2 and the corresponding iteration curves of the proposed method are given in Figure 6. The results show that the proposed method can achieve the desired stiffness accurately and efficiently. It is effective for the SCTD problem of the 3DJM.

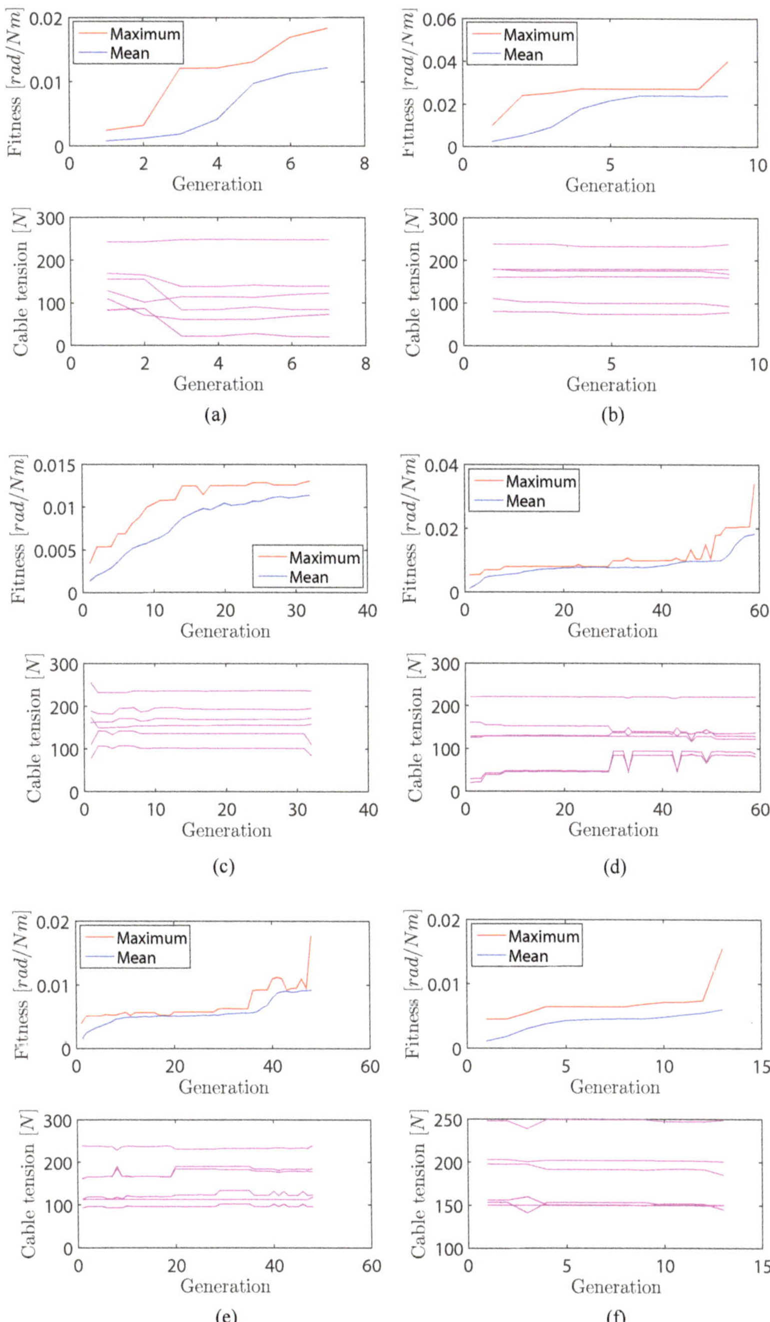

Figure 6. Iteration curves of the proposed method for the six sub-cases: (**a**) Case 1a, (**b**) Case 1b, (**c**) Case 1c, (**d**) Case 2a, (**e**) Case 2b, (**f**) Case 2c.

Table 2. Results of the simulation for the six sub-cases.

Case	Desired Stiffness K_{des}	Actual Stiffness $\overline{K}_{act}$ [Nm/rad]			$\eta(\overline{K}_{act}, K_{des})$
Case 1a	K_{des1a}	$\overline{K}_{act1a} = \begin{pmatrix} 1842.37 & 339.16 & 787.59 \\ 324.04 & 388.59 & -29.52 \\ 810.48 & -61.30 & 640.20 \end{pmatrix}$			99.9554%
Case 1b	K_{des1b}	$\overline{K}_{act1b} = \begin{pmatrix} 1115.72 & 171.67 & 42.79 \\ 155.02 & 1900.10 & -124.34 \\ 75.13 & -148.55 & 583.72 \end{pmatrix}$			99.9991%
Case 1c	K_{des1c}	$\overline{K}_{act1a} = \begin{pmatrix} 1593.57 & -345.80 & 335.65 \\ 335.65 & -357.47 & 1610.93 \\ 264.71 & 353.79 & 466.67 \end{pmatrix}$			99.9584%
Case 2a	K_{des2a}	$\overline{K}_{act2a} = \begin{pmatrix} 1316.56 & -106.95 & 503.13 \\ -109.93 & 636.37 & 137.96 \\ 524.24 & 128.03 & 504.65 \end{pmatrix}$			99.9500%
Case 2b	K_{des2b}	$\overline{K}_{act2b} = \begin{pmatrix} 1679.74 & 187.54 & 464.93 \\ 208.97 & 1276.25 & -366.06 \\ 489.22 & -343.72 & 676.79 \end{pmatrix}$			99.9697%
Case 2c	K_{des2c}	$\overline{K}_{act2c} = \begin{pmatrix} 1627.92 & 283.30 & 33.88 \\ 295.86 & 2390.02 & -270.50 \\ 73.42 & -282.15 & 710.07 \end{pmatrix}$			99.9629%

6. Conclusions

Inspired by the structure of human arms, a modular *cable-driven human-like robotic arm* (CHRA) was developed for safe human–robot interaction, since it has advantage of flexibility, low moving mass and intrinsic safety. Due to the unilateral driving properties of the cables, the CHRA is redundantly actuated and its stiffness can be adjusted by regulating the cable tensions. The cable tension distribution problem becomes a key element for the stiffness adjustment of the CHRA. Since the trajectory of the *3-DOF joint module* (3DJM) of the CHRA is a curve on Lie group SO(3), the stiffness of the 3DJM was evaluated by the covariant derivative of the load to the displacement on SO(3), called an enhanced stiffness model of the 3DJM. In this paper, we focus on analyzing how the cable tension distribution problem oriented the enhanced stiffness of the 3DJM. Since the enhanced stiffness model of the 3DJM is too complicated for solving the cable tensions from the desired stiffness analytically, the SCTD problem was formulated as a nonlinear optimization problem. By analyzing the rank of the Jacobian matrix and the equilibrium equation of the 3DJM, a variable elimination technique was employed to simplify the optimization model. A method based on the genetic algorithm was proposed to solve the optimization model, which only utilized the objective function values. The results of a comprehensive simulation show that the proposed method can solve the cable tension distribution from the desired stiffness accurately and efficiently.

Author Contributions: Conceptualization, K.Y. and G.Y.; supervision, G.Y.; project administration, C.Z. and Y.C.; writing—review and editing, C.C., T.Z. and T.C. All authors have read and agreed to the published version of the manuscript.

Funding: This research is funded by the Public Welfare Technology Research Program of Zhejiang Province, China (Grant number: LGF19E050001), National Natural Science Foundation of China (Grant number:51705510), National-Zhejiang Joint Natural Science Foundation of China (Grant number: U1909215), Institute of robotics and intelligent manufacturing innovation, Chinese Academy of Science (Grant number: C2018005), Ningbo Natural Science Foundation (Grant number: 2019A610114), Open Research Project of the State Key Laboratory of Industrial Control Technology, Zhejiang University, China (Grant number: ICT20048) and Public Welfare Technology Research Program of Zhoushan, China (Grant number: 2019C31067).

Conflicts of Interest: The authors declare no conflict of interest. The funders had no role in the design of the study; in the collection, analyses, or interpretation of data; in the writing of the manuscript, or in the decision to publish the results.

Appendix A. Analysis of the Rank of the Jacobian Matrix of the 3DJM

The transpose of the Jacobian matrix $J \in \mathbb{R}^{6\times3}$ is denoted as $S \in \mathbb{R}^{3\times6}$, which is represented as below

$$S = J^T = -(a_1 \times u_1, a_2 \times u_2, \cdots, a_6 \times u_6)^T. \tag{A1}$$

Assuming the column rank of S satisfies

$$\mathbf{rank}(S) < 3, \tag{A2}$$

the mixed product of any three column vectors of the matrix S yields

$$S_i \cdot (S_j \times S_k) = \begin{vmatrix} s_{i1} & s_{j1} & s_{k1} \\ s_{i2} & s_{j2} & s_{k2} \\ s_{i3} & s_{j3} & s_{k3} \end{vmatrix} \equiv 0, \tag{A3}$$

where $i, j, k = 1, 2, \ldots, 6$.

For the 3DJM of the CHRA, the mixed product of any three column vectors of the matrix S can be written as

$$\begin{aligned} &S_i \cdot (S_j \times S_k) \\ =&(a_i \times u_i) \cdot [(a_j \times u_j) \times (a_k \times u_k)] \\ =&\frac{[a_i \times (b_i - a_i)] \cdot \{[a_j \times (b_j - a_j)] \times [a_k \times (b_k - a_k)]\}}{c_i c_j c_k} \\ =&\frac{(a_i \times b_i) \cdot [(a_j \times b_j) \times (a_k \times b_k)]}{c_i c_j c_k} \\ \equiv&0 \quad (i, j, k = 1, 2, \cdots, 6). \end{aligned} \tag{A4}$$

In the design of the 3DJM, $e_A \approx 0$ and $e_B \approx 0$, we have $a_1 \approx a_2, a_3 \approx a_4, a_5 \approx a_6$, and $b_1 \approx b_6, b_2 \approx b_3, b_4 \approx b_5$. Then, the following mixed product can be represented as

$$\begin{aligned} &(a_1 \times b_1) \cdot [(a_2 \times b_2) \times (a_3 \times b_3)] \\ \approx&(a_1 \times b_1) \cdot [(a_1 \times b_3) \times (a_3 \times b_3)] \\ =&(a_1 \times b_1) \cdot (\mu b_3) \\ =&0, \end{aligned} \tag{A5}$$

$$\begin{aligned} &(a_4 \times b_4) \cdot [(a_2 \times b_2) \times (a_3 \times b_3)] \\ \approx&(a_3 \times b_5) \cdot [(a_1 \times b_3) \times (a_3 \times b_3)] \\ =&(a_3 \times b_5) \cdot (\mu b_3) \\ \equiv&0, \end{aligned} \tag{A6}$$

$$\begin{aligned} &(a_5 \times b_5) \cdot [(a_2 \times b_2) \times (a_3 \times b_3)] \\ \approx&(a_5 \times b_5) \cdot [(a_1 \times b_3) \times (a_3 \times b_3)] \\ =&(a_5 \times b_5) \cdot (\mu b_3) \\ \equiv&0, \end{aligned} \tag{A7}$$

where $\mu \in \mathbb{R}$ and $\mu \neq 0$.

Equation (A5) shows that vectors b_3, a_1 and b_1 are in the same plane, (A6) shows that vectors b_3, a_3 and b_5 are in the same plane, and (A7) shows that vectors b_3, a_5 and b_5 are in the same plane.

In summary, we have that vectors a_1, a_3, a_5, b_1, b_3 and b_5 are in the same plane, which is impossible for the 3DJM. The assumption (A2) does not hold, hence

$$\mathbf{rank}(S) = 3, \tag{A8}$$

i.e.,

$$\mathbf{rank}(J) = 3. \tag{A9}$$

References

1. Giorelli, M.; Renda, F.; Calisti, M.; Arienti, A.; Laschi, C. Learning the inverse kinetics of an octopus-like manipulator in three-dimensional space. *BioInspir. Biomimetics* **2015**, *10*, 035006. [CrossRef] [PubMed]
2. Lau, D.; Oetomo, D.; Halgamuge, S.K. Inverse Dynamics of Multilink Cable-Driven Manipulators with the Consideration of Joint Interaction Forces and Moments. *IEEE Trans. Robot.* **2015**, *31*, 479–488. [CrossRef]
3. Lau, D.; Eden, J.; Oetomo, D.; Halgamuge, S.K. Musculoskeletal StaticWorkspace Analysis of the Human Shoulder as a Cable-Driven Robot. *IEEE Trans. Mechatron.* **2015**, *20*, 978–984. [CrossRef]
4. Wang, Y.; Song, C.; Zheng, T.; Lau, D.; Yang, K.; Yang, G. Cable Routing Design and Performance Evaluation for Multi-link Cable-Driven Robots with Minimal Number of Actuating Cables. *IEEE Access* **2019**, *7*, 135790–135800. [CrossRef]
5. Wang, H.; Zhang, R.; Chen, W.; Wang, X.; Pfeifer, R. A cable-driven soft robot surgical system for cardiothoracic endoscopic surgery-preclinical tests in animals. *Surg. Endosc.* **2017**, *31*, 3153–3158. [CrossRef]
6. Li, J.; Lam, J.; Liu, M.; Wang, Z. Compliant Control and Compensation for A Compact Cable-Driven Robotic Manipulator. *IEEE Robot. Autom. Lett.* **2020**, *5*, 5417–5424. [CrossRef]
7. Hwang, M.; Thananjeyan, B.; Paradis, S.; Seita, D.; Ichnowski, J.; Fer, D.; Low, T.; Goldberg, K. Efficiently Calibrating Cable-Driven Surgical Robots With RGBD Fiducial Sensing and Recurrent Neural Networks. *IEEE Robot. Autom. Lett.* **2020**, *5*, 5937–5944. [CrossRef]
8. Cui, X.; Chen, W.; Jin, X.; Agrawal, S.K. Design of a 7-DOF cable-driven arm exoskeleton (CAREX-7) and a controller for dexterous motion training or assistance. *IEEE/ASME Trans. Mechatron.* **2017**, *22*, 161–172. [CrossRef]
9. Kuan, J.; Pasch, K.A.; Herr, H.M. A high-performance cable-drive module for the development of wearable devices. *IEEE/ASME Trans. Mechatron.* **2018**, *23*, 1238–1248. [CrossRef]
10. Chen, C.T.; Lien, W.Y.; Chen, C.T.; Twu, M.J.; Wu, Y.C. Dynamic Modeling and Motion Control of a Cable-Driven Robotic Exoskeleton With Pneumatic Artificial Muscle Actuators. *IEEE Access* **2020**, *8*, 149796–149807. [CrossRef]
11. Samper-Escudero, J.L.; Gimenez-Fernandez, A.; Sanchez-Uran, M.A.; Frerre, M. A Cable-Driven Exosuit for Upper Limb Flexion Based on Fibres Compliance. *IEEE Access* **2020**, *8*, 153297–153310. [CrossRef]
12. Chen, T.; Casas, R.; Lum, P.S. An Elbow Exoskeleton for Upper Limb Rehabilitation With Series Elastic Actuator and Cable-Driven Differential. *IEEE Trans. Robot.* **2019**, *35*, 1464–1474. [CrossRef] [PubMed]
13. Dong, X.; Axinte, D.; Palmer, D.; Cobos, S.; Raffles, M.; Rabani, A.; Kell, J. Development of a slender continuum robotic system for on-wing inspection/repair of gas turbine engines. *Robot. Comput. Integr. Manuf.* **2017**, *44*, 218–229. [CrossRef]
14. Wang, M.; Palmer, D.; Dong, X.; Alatorre, D.; Axinte, D.; Norton, A. Design and Development of a Slender Dual-Structure Continuum Robot for In-Situ Aeroengine Repair. In Proceedings of the IEEE/RSJ International Conference on Intelligent Robots and Systems (IROS), Madrid, Spain, 1–5 October 2018; pp. 5648–5653.
15. Axinte, D.; Dong, X.; Palmer, D.; Rushworth, A.; Guzman, S.C.; Olarra, A.; Arizaga, I.; Gomez-Acedo, E.; Txoperena, K.; Pfeiffer, K.; et al. MiRoR-Miniaturized Robotic Systems for HolisticIn-SituRepair and Maintenance Works in Restrained and Hazardous Environments. *IEEE/ASME Trans. Mechatron.* **2018**, *23*, 978–981. [CrossRef]
16. Campeau-Lecours, A.; Foucault, S.; Laliberte, T.; Mayer-St-Onge, B.; Gosselin, C. A cable-suspended intelligent crane assist device for the intuitive manipulation of large payloads. *IEEE/ASME Trans. Mechatron.* **2016**, *21*, 2073–2084. [CrossRef]

17. Dion-Gauvin, P.; Gosselin, C. Dynamic point-to-point trajectory planning of a three-DOF cable-suspended mechanism using the hypocycloid curve. *IEEE/ASME Trans. Mechatron.* **2018**, *23*, 1964–1972. [CrossRef]

18. Wang, H.; Kinugawa, J.; Kosuge, K. Exact kinematic modeling and identification of reconfigurable cable-driven robots with dual-pulley cable guiding mechanisms. *IEEE/ASME Trans. Mechatron.* **2019**, *24*, 774–784. [CrossRef]

19. Zhang, F.; Shang, W.; Zhang, B.; Cong, S. Design Optimization of Redundantly Actuated Cable-Driven Parallel Robots for Automated Warehouse System. *IEEE Access* **2020**, *8*, 56867–56879. [CrossRef]

20. Oh, S.R.; Agrawal, S.K. Cable Suspended Planar Robots with Redundant Cables: Controllers with Positive Tensions. *IEEE Trans. Robot.* **2005**, *21*, 457–465.

21. Bruckmann, T.; Pott, A.; Hiller, M. Calculating force distributions for redundantly actuated tendon-based Stewart platforms. In *Advances in Robot Kinematics*; Springer: Berlin/Heidelberg, Germany, 2006; pp. 403–412.

22. Hassan, M.; Khajepour, A. Optimization of Actuator Forces in Cable-Based Parallel Manipulators Using Convex Analysis. *IEEE Trans. Robot.* **2008**, *24*, 736–740. [CrossRef]

23. Borgstrom, P.H.; Jordan, B.L.; Borgstrom, B.J.; Stealey, M.J.; Sukhatme, G.S.; Batalin, M.A.; Kaiser, W.J. NIMS-PL: A Cable-driven Robot with Self-calibration Capabilities. *IEEE Trans. Robot.* **2009**, *25*, 1005–1015. [CrossRef]

24. You, X.; Bing, L.; Chen, W.; Yu, S. Tension distribution algorithm of a 7-DOF cable-driven robotic arm based on dynamic minimum pre-tightening force. In Proceedings of the IEEE International Conference on Robotics and Biomimetics (ROBIO), Karon Beach, Thailand, 7–11 December 2011; pp. 715–720.

25. Cote, A.F.; Cardou, P.; Gosselin, C. A tension distribution algorithm for cable-driven parallel robots operating beyond their wrench-feasible workspace. In Proceedings of the International Conference on Control, Automation and Systems, Jeju, Korea, 18–21 October 2017; pp. 68–73.

26. Fang, S.; Franitza, D.; Torlo, M.; Bekes, F.; Hiller, M. Motion Control of a Tendon-BasedParallel Manipulator Using Optimal Tension Distribution. *IEEE/ASME Trans. Mechatron.* **2004**, *9*, 561–568. [CrossRef]

27. Gosselin, C. On the Determination of the Force Distribution in Overconstrained Cable-driven Parallel Mechanisms. *Meccanica* **2011**, *46*, 3–15. [CrossRef]

28. Lim, W.B.; Song, H.Y.; Yang, G. Optimization of Tension Distribution for Cable-Driven Manipulators Using Tension-Level Index. *IEEE/ASME Trans. Mechatron.* **2014**, *19*, 676–683. [CrossRef]

29. Yang, K.; Yang, G.; Chen, S.L.; Wang, Y.; Zhang, C.; Fang, Z.; Zheng, T.; Wang, C. Study on Stiffness-Oriented Cable Tension Distribution for a Symmetrical Cable-Driven Mechanism. *Symmetry* **2019**, *11*, 1158. [CrossRef]

30. Yang, K.; Yang, G.; Chen, S.L.; Wang, Y.; Zheng, T.; Fang, Z.; Wang, C. Enhanced Stiffness Modeling and Identification Method for a Cable-driven Spherical Joint Module. *IEEE Access* **2019**, *7*, 137875–137886. [CrossRef]

31. Murray, R.N.; Li, Z.; Sastry, S. *A Mathematical Introduction to Robotics Manipulation*; CRC Press: Boca Raton, FL, USA, 1994.

32. Lee, J.M. *Introduction to Riemannian Manifolds*; Springer: Berlin/Heidelberg, Germany, 2018.

33. Zefran, M.; Kumar, V. A geometric approach to the study of the Cartesian stiffness matrix. *J. Mech. Des.* **2002**, *124*, 30–38. [CrossRef]

34. Golub, G.H.; Charles, F.V.L. *Matrix Computations*; The Johns Hopkins University Press: Baltimore, MD, USA, 1985.

35. Nelder, J.A.; Mead, R. A simplex method for function minimization. *Comput. J.* **1965**, *7*, 308–313. [CrossRef]

36. Affenzeller, M.; Winkler, S.; Wagner, S.; Beham, A. *Genetic Algorithms and Genetic Programming-Modern Concepts and Practical Applications*; CRC Press: Boca Raton, FL, USA, 2009.

Publisher's Note: MDPI stays neutral with regard to jurisdictional claims in published maps and institutional affiliations.

applied sciences

Article

A Novel Type of Wall-Climbing Robot with a Gear Transmission System Arm and Adhere Mechanism Inspired by Cicada and Gecko

Shiyuan Bian [1,2], Feng Xu [1,2], Yuliang Wei [1,2] and Deyi Kong [1,3,*]

1 Institute of Intelligent Machines, Hefei Institutes of Physical Science, Chinese Academy of Sciences, Hefei 230031, China; sybian@iim.ac.cn (S.B.); xf1992@mail.ustc.edu.cn (F.X.); wyl1010@mail.ustc.edu.cn (Y.W.)
2 Science Island Branch of Graduate School, University of Science and Technology of China, Hefei 230009, China
3 Innovation Academy for Seed Design, Chinese Academy of Sciences, Sanya 572024, China
* Correspondence: kongdy@iim.ac.cn

Abstract: To support the inspections of different contact walls (rough and smooth), a novel type of wall-climbing robot was proposed. Its design embodied a new gear transmission system arm and an adherence mechanism inspired by cicadas and geckos. The actuating structure consisted of a five-bar link and a gear transmission for the arm stretching, which was driven by the servos. The linkers and gears formed the palm of this robot for climbing on a line. Moreover, the robot's adherence method for the rough surfaces used bionic spine materials inspired by the cicada. For smooth surface, a bionic adhesion material was proposed inspired by the gecko. To assess the adherence mechanism of the cicada and gecko, the electron microscope images of the palm of the cicada and gecko were obtained by an electron microscope. The 3D printing technology and photolithography technology were utilized to manufacture the robot's structures. The adherence force experiments demonstrated the bionic spines and bionic materials achieved good climbing on cloth, stones, and glass surfaces. Furthermore, a new gait for the robot was designed to ensure its stability. The dynamic characteristics of the robot's gear transmission were obtained.

Keywords: wall-climbing robot; gear transmission; bionic spine; electron microscope images; 3D printing technology

check for **updates**

Citation: Bian, S.; Xu, F.; Wei, Y.; Kong, D. A Novel Type of Wall-Climbing Robot with a Gear Transmission System Arm and Adhere Mechanism Inspired by Cicada and Gecko. *Appl. Sci.* **2021**, *11*, 4137. https://doi.org/10.3390/app11094137

Academic Editor: TaeWon Seo

Received: 12 March 2021
Accepted: 28 April 2021
Published: 30 April 2021

Publisher's Note: MDPI stays neutral with regard to jurisdictional claims in published maps and institutional affiliations.

1. Introduction

Researchers investigating wall-climbing robots looked to nature for inspiration. Around the gecko, researchers studied the gecko's toe and crawling approach to propose some wall-climbing robots. On the basis of numerous studies on the gecko [1–6], the gecko's adhesion mechanism was proposed, and the relevant wall-climbing robots inspired by the gecko were investigated. Metin et al. [7] assessed a four-legged wall-climbing robot with an elastomer adhesive pad to climb a sloped surface. A novel robot [8] was presented with a shape memory alloy actuator to drive its climbing. A microfiber with an angled mushroom tips [9] array inspired by gecko was manufactured for a novel wall-climbing robot, Waalbot [10,11]. These studies applied the special structure array inspired by the gecko to the wall-climbing robot. Cutkosky et al. [12–15] learned from the gecko's pad and body structure to present a gecko-like, wall-climbing robot. A bionic pad of this robot consisted of a microfiber array and four-legged locomotion. The tail was inspired by the gecko's toe and the gecko's climbing gait. Dai et al. [16] proposed a novel wall-climbing robot with a biomimetic gait inspired by the gecko, which adjusted the posture of the robot's foot to stably adhere to the wall. Mei et al. [17,18] assessed a tank-like, wall-climbing robot with gecko-inspired material on the caterpillar band. This robot climbed on a vertical surface, on the ceiling, or travelled from the vertical surface to the ceiling.

85

In another line of research, a series of studies were presented on wall-climbing robots inspired by animals with claws for traversing rough surfaces. Dai et al. [19] proposed a model of claws' adherence on rough surfaces and investigated the relation of the dimensions of the claws with respect to the particles. Cutkosky et al. [20–22] presented a six-legged, wall-climbing robot with alternating tripod gait for climbing a wall with stability. Saunders et al. [23] proposed a RiSE robot with six legs and flexible spines inspired by insects. The robot climbed a 90° tree in addition to carpeted and cork covered stucco surfaces in the quasi-static regime. Then, the foot of this robot was improved to climb on the wall with stability. A new type of the wall-climbing robot [24] was assessed and demonstrated better performance. An improved wall-climbing robot [25] was proposed with better stability and faster movement. Based on the results presented by Goldman et al. [26] on the behaviors of the cockroach, Clark et al. [27–30] proposed a bionic, double-legged wall-climbing robot for rough surfaces. Furthermore, quadrupedal wall-climbing robots [31–33] were proposed with a lateral leg spring model. Liu [34] proposed a wheeled wall-climbing robot with bionic feet inspired by Serica orientalis Motschulsky; it was capable of climbing across the transition from horizontal to vertical surfaces. Then, a novel, tracked, wall-climbing robot with bio-inspired spine feet was presented for rough surface [35].

It can be seen from the literature that the wall-climbing robots studied by many researchers only aim at smooth or rough surfaces and rarely prepare wall-climbing robots that adapt to both smooth and rough surfaces. This paper proposes a novel type of wall-climbing robot with a gear transmission system arm for climbing on the smooth and rough surfaces that absorbs the adhesion characteristics of gecko and cicada. The novel actuating structure consisted of a five-bar link, gear transmission system, and drive source servos. The dynamic characteristics of the gear transmission of the robot were obtained by a numerical model. In the transmission system, the driving gear was driven by the servos, and the two driven gears were fixed with the links. In this approach, inspired by the cicada and gecko, the palm of the robot conformed to the line when climbing. Moreover, the bionic spine adherence materials and gecko-inspired materials were used to traverse rough surfaces and smooth surfaces. The robots' bodies were manufactured by using the 3D printing technology with resin. The adherence force experiments of the bionic spines and gecko-inspired materials were achieved on the rough and smooth surfaces. Finally, a new gait of the robot was designed to ensure the robot's climbing stability. The experiments of the robot climbing on the different walls are presented at the end. Based on the study of the bionic hooks and the materials bio-inspired by gecko, this robot, which can adapt to rough and smooth wall surfaces, was put forward innovatively.

2. Modeling of Wall-Climbing Robot

The cicada and gecko can climb on walls because of their feet with special microstructures. These microstructures give support for adhesion force for the cicada and gecko to adhere to a tree. This special adhesion ability was adopted in this robot for crawling on the wall with stability.

2.1. Robot Description

Figure 1a presents a novel, wall-climbing robot whose body was inspired by a gecko, with a gear transmission system arm. The robot included a head, a main body fixed on the control element and power module, feet driven by a gear transmission fixed on a five-bar link mechanism, and tail for supporting the robot climbing on a wall. The tail was connected to the body by a torsional ring, which provided a support force to the robot during climb. The five-bar link is presented in Figure 1b. The driving wheel is gear 1; gear 2 and gear 3 are the driven wheels. The link 1 and link 4 were fixed on gear 2 and gear 3, respectively. The four links (link 1, link 2, link 3, link 4) were connected by small bearings. The driving wheel gear 1 meshed with gear 2 and gear 3 at the same time on the supporting part to keep the palm moving on the line. Figure 1c presents the prototype of this robot designed by 3D printing technology with resin material. The bio-inspired hooks were assembled to the palm. The length of a single hook was 18 mm. The length of the body was 474 mm (including tail), and the body width was 315 mm (including the four legs).

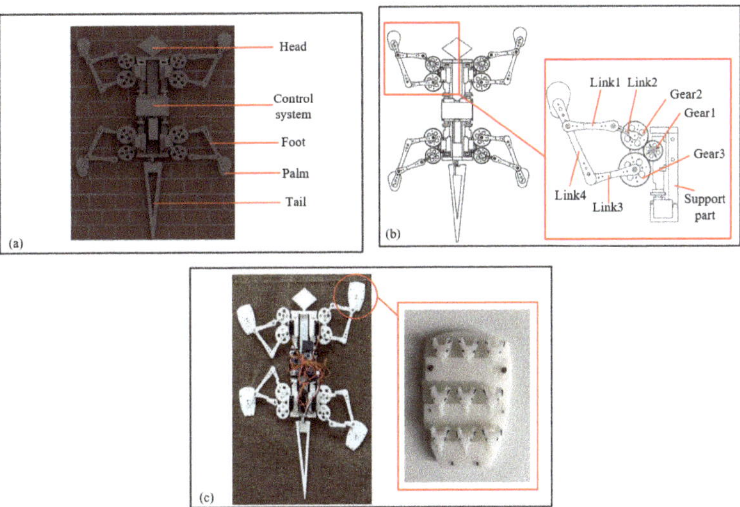

Figure 1. The design of a bio-inspired, wall-climbing robot: (**a**) the components of this robot, (**b**) the mechanism structure of the foot on the robot, (**c**) the prototype of this robot.

2.2. Statics' Analysis

This robot, adhering on the wall by its four palms and the tail, is shown in Figure 2a. The palms and tail provided the force to prevent the robot from falling down and capsizing.

$$\begin{cases} \sum_{i=1}^{4} F_{iy} + F_{ty} = G \\ \sum_{i=1}^{4} F_{ix}l_i + F_{tx}l_t = Gh \end{cases} \tag{1}$$

Here, G is the weight of the robot, F_{iy} ($i = 1, 2, 3, 4$) is the component of the force of the palms on the vertical destination, and F_{ty} is the component of the force of the tail on the vertical destination.

Figure 2b shows the force analysis of the robot's palm lift from B1 to B2 away from the surface by the servo located horizontally. The servo provides the momentum to lift the palm away from the surface, and the force analysis is shown as:

$$(G_p + F_a)L_p \cos(\theta_p) = M_p \tag{2}$$

Here, G_p represents the palm's weight, F_a is the adherence force of the palm lift from the surface by the bionic adherence material, L_p represents the length of the center position of the palm to the rotation center of the servo position O, θ_p is the angle the palm rotates from the position B1 to B2, and M_p is the momentum of the servo on the part.

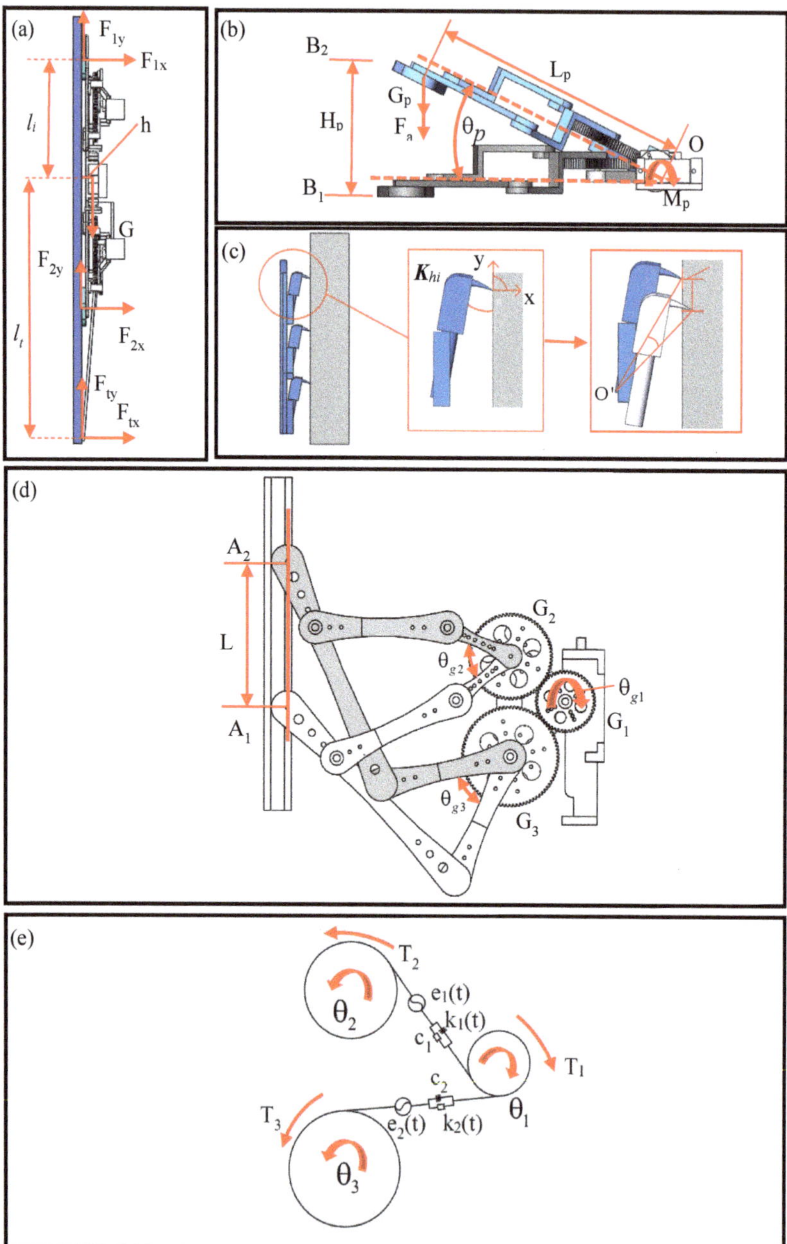

Figure 2. (**a**) Force analysis of the wall-climbing robot climbing on the wall, (**b**) force analysis of the robot's palm lifted to detach off the surface, (**c**) force analysis of the robot's palm adhere to the surface, (**d**) the motion analysis of the robot's palm for climbing, (**e**) the dynamic mechanics' mode of the transmission system.

The palm of this robot adhering to the wall is shown in Figure 2c. To investigate the adhesion force of the hooks, the stiffness of the hooks is K_h, the rotation angle is defined as θ', and the adhesion force of the hooks is

$$F_h = K_h \delta_h \tag{3}$$

Here, the K_h is the stiffness matrix and δ_h is the deflection of the hook.

$$F_h = \begin{pmatrix} F_{hx} \\ F_{hy} \\ M_{h\theta'} \end{pmatrix}, \; \delta_h = \begin{pmatrix} \Delta_x \\ \Delta_y \\ \Delta_{\theta'} \end{pmatrix} \tag{4}$$

Here, the F_{hx} is the force of the wall on the hook in the x direction, the F_{hy} is the force in direction y, and $M_{h\theta'}$ is the momentum of the hook. The Δ_x, Δ_y, and $\Delta_{\theta'}$ are the deflection of the hook on the wall in x, y, and bend directions.

The servos provided the torque to drive gear 1 (G1); in turn, gear 2 and gear 3 were driven. Meanwhile, link 1 and link 4 were fixed on the gears. They were driven by gear 2 and gear 3 for stretching to crawl on the wall. Figure 2d represents the two positions of the robot's palm driven by a servo through the five-link mechanism (A1, A2). The end of the link moved to the symmetry position (A1–A2) to keep the robot's palm moving in a straight line. While the robot's palm stretched to position A2, gear 1, gear 2, and gear 3 turned θ_{g1}, θ_{g2} and θ_{g3}, respectively. During this situation, the linear velocity of the gears was kept the same:

$$V_1(t) = V_2(t) = V_3(t) \tag{5}$$

hence,

$$\omega_1(t)r_1 = \omega_2(t)r_2 = \omega_3(t)r_3 \tag{6}$$

$$\theta_{g1}(t) = \omega t \tag{7}$$

$$\theta_{g1}(t)d_1 = \theta_{g2}(t)d_2 = \theta_{g3}(t)d_3 \tag{8}$$

$$d = mz \tag{9}$$

$$\theta_{g1}(t)z_1 = \theta_{g2}(t)z_2 = \theta_{g3}(t)z_3 \tag{10}$$

Here, V_1, V_2, and V_3 are the linear velocity of the gears; ω_1, ω_2, ω_3 represent the angular velocities of the gears; d shows the pitch diameter of the gear; m is the module of the gear; z is the number of teeth; and z_1, z_2, and z_3 represent the number of teeth of the gear 1, gear 2, and gear 3, respectively. The parameters of gear 1, gear 2, and gear 3 can be obtained by Equation (10), based on the length of the links.

2.3. Kinetic Analysis

This robot had four legs for climbing on the wall. The legs were driven by the gear transmission. The gear transmission affected the stability of crawling system. Hence, the kinetic analysis of the gear transmission was so important that the relevant dynamic model was proposed. Before the model building, the following hypothesis was proposed:

1. The internal and external mesh pairs were represented by virtual spring-damping units, time-varying stiffness, and comprehensive transmission error.
2. The frictions of each meshing pair and supporting bearings could be neglected.
3. The backlashes of gear pairs were included.

Figure 2e presents the dynamic mechanics of the gear transmission system in a foot of this robot. The θ_i ($I = 1, 2, 3$) is the displacement of the gears, I_i ($i = 1, 2, 3$) represents the moment of inertia of the gear, and e_j (t), ($j = 1, 2$) is the comprehensive transmission error during the process of gear driving. The comprehensive transmission error e_j (t) is periodic and varies with the mesh frequency ω_j, which can be written as

$$e_j(t) = E_j \cos(\omega_j t + \varphi_j) \tag{11}$$

where E_j is the comprehensive gear meshing error amplitude and φ_j is the phase angle of the j_{th} gear pair. The $k_j(t)$, $(j = 1, 2)$ is the time-varying mesh stiffness and can be described as

$$k_j(t) = k_{mj} + k_s \cos(\omega_j t + \varphi_{kj}) \tag{12}$$

where k_{mj} is the average amplitude, k_s is the varying amplitude of the j_{th} time-varying mesh stiffness of gear pair, φ_{kj} is the phase angle of this gear pair, c_j, $(j = 1, 2)$ is the damping coefficient, T_1 is the driving moment, and T_2 and T_3 are the driven moments on the gear. The dynamic mechanics' function of this gear system can be written as

$$\begin{cases} I_2\ddot{\theta}_2 + c_1 R_2(R_2\dot{\theta}_2 - R_1\dot{\theta}_1 - \dot{e}_1(t)) + k_1(t)R_2 f(R_2\theta_2 - R_1\theta_1 - e_1(t)) = T_2 \\[2mm] I_3\ddot{\theta}_3 + c_2 R_3(R_3\dot{\theta}_3 - R_1\dot{\theta}_1 - \dot{e}_2(t)) + k_2(t)R_3 f(R_3\theta_3 - R_1\theta_1 - e_2(t)) = T_3 \\[2mm] I_1\ddot{\theta}_1 - c_1 R_1(R_2\dot{\theta}_2 - R_1\dot{\theta}_1 - \dot{e}_1(t)) - k_1(t)R_1 f(R_2\theta_2 - R_1\theta_1 - e_1(t)) \\[2mm] \quad -c_2 R_1(R_3\dot{\theta}_3 - R_1\dot{\theta}_1 - \dot{e}_2(t)) - k_2(t)R_1 f(R_3\theta_3 - R_1\theta_1 - e_2(t)) = -T_1 \end{cases} \tag{13}$$

Here, R_i $(i = 1, 2, 3)$ represents the radius of the gears. This model only included the torsional direction displacement. This gear transmission system included the gear backlash, represented by Figure 2e. The displacement function $f(X_i)$ that included the gear backlash can be expressed as

$$f(X_j) = \begin{cases} X_j - b & X_j > b \\ 0 & |X_j| \leq b \\ X_j + b & X_j < -b \end{cases} \tag{14}$$

In order to eliminate the rigid body movement, the relative displacements of the gear pair were proposed as X_j $(j = 1, 2)$. These can be obtained as

$$\begin{cases} X_1 = R_2\theta_2 - R_1\theta_1 - e_1(t) \\[2mm] X_2 = R_3\theta_3 - R_1\theta_1 - e_2(t) \end{cases} \tag{15}$$

Hence, the function can be written

$$\begin{cases} \ddot{X}_1 + c_1 R_2^2/I_2 \dot{X}_1 + c_1 R_1^2/I_1 \dot{X}_1 + c_2 R_1^2/I_1 \dot{X}_2 + k_1(t)R_2^2/I_2 f(X_1) + k_1(t)R_1^2/I_1 f(X_1) \\[2mm] \quad + k_2(t)R_1^2/I_1 f(X_2) = -\ddot{e}_1(t) + R_2 T_2/I_2 + R_1 T_1/I_1 \\[2mm] \ddot{X}_2 + c_2 R_3^2/I_3 \dot{X}_2 + c_2 R_1^2/I_1 \dot{X}_2 + c_1 R_1^2/I_1 \dot{X}_1 + k_2(t)R_3^2/I_3 f(X_2) + k_2(t)R_1^2/I_1 f(X_2) \\[2mm] \quad + k_1(t)R_1^2/I_1 f(X_1) = -\ddot{e}_2(t) + R_1 T_1/I_1 + R_3 T_3/I_3 \end{cases} \tag{16}$$

The dimensionless process was used to solve this function to obtain the vibration placement. The normal dimension b_c is

$$b_c = b/\bar{b} \tag{17}$$

where the $\bar{b}$ is dimension value. The characteristic frequency ω_n can be expressed as

$$\omega_n = \sqrt{k_{mj}/(1/m_1 + 1/m_2)} \tag{18}$$

The dimensionless treatment of displacement, velocity, and acceleration are shown as

$$\begin{cases} X_i = \overline{X}_i b_c \\ \dot{X}_i = \dot{\overline{X}}_i \omega_n b_c \\ \ddot{X}_i = \ddot{\overline{X}}_i \omega_n^2 b_c \end{cases} \tag{19}$$

The gear backlash can be expressed as

$$f(\overline{X}_j) = \begin{cases} \overline{X}_j - \overline{b} & \overline{X}_j > \overline{b} \\ 0 & |\overline{X}_j| \leq \overline{b} \\ \overline{X}_j + \overline{b} & \overline{X}_j < -\overline{b} \end{cases} \tag{20}$$

The time-varying mesh stiffness and comprehensive transmission error can be presented as

$$\begin{cases} \overline{e}_j(t) = \overline{E}_j \cos(\overline{\omega}_j t + \varphi_j) \\ \overline{k}_j(t) = k_{mj} + k_s \cos(\overline{\omega}_j t + \varphi_{kj}) \end{cases} \tag{21}$$

where $\overline{\omega}_j = \omega_j/\omega_n$, $\overline{E}_j = E_j/b_c$. The function can be expressed as

$$\begin{cases} \omega_n^2 b_c \ddot{\overline{X}}_1 + \omega_n b_c c_1 R_2^2/I_2 \dot{\overline{X}}_1 + \omega_n b_c c_1 R_1^2/I_1 \dot{\overline{X}}_1 + \omega_n b_c c_2 R_1^2 I_2 \dot{\overline{X}}_2 + b_c k_1(t) R_2^2/I_2 f(\overline{X}_1) \\ \qquad + b_c k_1(t) R_1^2/I_1 f(\overline{X}_1) + b_c k_2(t) R_1^2/I_1 f(\overline{X}_2) = -\ddot{\overline{e}}_1(t) + R_2 T_2/I_2 + R_1 T_1/I_1 \\ \\ \omega_n^2 b_c \ddot{\overline{X}}_2 + \omega_n b_c c_2 R_3^2/I_3 \dot{\overline{X}}_2 + \omega_n b_c c_2 R_1^2/I_1 \dot{\overline{X}}_2 + \omega_n b_c c_1 R_1^2 I_3 \dot{\overline{X}}_1 + b_c k_2(t) R_3^2/I_3 f(\overline{X}_2) \\ \qquad + b_c k_2(t) R_1^2/I_1 f(\overline{X}_2) + b_c k_1(t) R_1^2/I_1 f(\overline{X}_1) = -\ddot{\overline{e}}_2(t) + R_1 T_1/I_1 + R_3 T_3/I_3 \end{cases} \tag{22}$$

$$\begin{cases} \ddot{\overline{X}}_1 = -\ddot{\overline{e}}_1(t) + T_2/(R_2 \omega_n^2 b_c m_2) + T_1/(R_1 \omega_n^2 b_c m_1) - (c_1/(\omega_n m_2) - c_1/(\omega_n m_1))\dot{\overline{X}}_1 \\ \qquad -c_2/(\omega_n m_1)\dot{\overline{X}}_2 - k_1(t)(1/(\omega_n^2 m_2) - 1/(\omega_n^2 m_1))f(\overline{X}_1) - k_2(t)/(\omega_n^2 m_1)f(\overline{X}_2) \\ \\ \ddot{\overline{X}}_2 = -\ddot{\overline{e}}_2(t) + T_1/(R_1 \omega_n^2 b_c m_1) + T_3/(R_3 \omega_n^2 b_c m_3) - (c_2/(\omega_n m_3) - c_2/(\omega_n m_1))\dot{\overline{X}}_2 \\ \qquad -c_1/(m_1 \omega_n)\dot{\overline{X}}_1 - k_2(t)(1/(\omega_n^2 m_3) - 1/(\omega_n^2 m_1))f(\overline{X}_2) - k_1(t)/(\omega_n^2 m_1)f(\overline{X}_1) \end{cases} \tag{23}$$

This can be written as

$$M\ddot{\overline{X}} + C\dot{\overline{X}} + K\overline{X} = \overline{F} \tag{24}$$

where

$$M = \begin{pmatrix} 1 & 0 \\ 0 & 1 \end{pmatrix}, \quad \overline{X} = \begin{pmatrix} \overline{X}_1 \\ \overline{X}_2 \end{pmatrix} \tag{25}$$

$$C = \begin{pmatrix} -c_1/\omega_n(1/m_2 + 1/m_1) & -c_2/(\omega_n m_1) \\ -c_1/(m_1 \omega_n) & -c_2/\omega_n(1/m_3 + 1/m_1) \end{pmatrix} \tag{26}$$

$$K = \begin{pmatrix} -k_1(t)(1/m_2 + 1/m_1)/\omega_n^2 & -k_2(t)/(\omega_n^2 m_1) \\ -k_1(t)/(m_1 \omega_n^2) & -k_2(t)(1/m_3 + 1/m_1)/\omega_n^2 \end{pmatrix} \tag{27}$$

$$F = \begin{pmatrix} T_2/(R_2 \omega_n^2 b_c m_2) + T_1/(R_1 \omega_n^2 b_c m_1) - \ddot{\overline{e}}_1(t) \\ T_3/(R_3 \omega_n^2 b_c m_3) + T_1/(R_1 \omega_n^2 b_c m_1) - \ddot{\overline{e}}_2(t) \end{pmatrix} \tag{28}$$

Equation (24) of the gear transmission system can be solved by the Runge–Kutta method.

The parameters of the whole system are shown in Table 1.

Table 1. The parameters of the gears.

Gear	Teeth	Module	Pressure Angle (°)	Width (mm)	Mass (g)
Gear 1	50	0.5	20	5	1.96
Gear 2	71	0.5	20	5	4.17
Gear 3	83	0.5	20	5	6

The relative displacements of gear 1, gear 2, and gear 3 were obtained from Equation (24) by the Runge–Kutta method. The results are shown in Figure 3. Figure 3a represents the dimensionless relative displacements of gear 1 and gear 2; these ranged between –0.1902~0.1905. Figure 3b represents the dimensionless relative displacements of gear 1 and gear 3; these ranged between −0.2001~0.20. The dimensionless relative displacements of gear 1 and gear 2 were larger than gear 1 and gear 3. The deformations of gear 1 and gear 3 were larger than that of gear 1 and gear 2. Therefore, the vibration displacement of gear 3 was larger than the gear 2. To investigate the stability of these three gears, the phase images and Poincare maps are presented in Figure 4. From Figure 4, the phase images of gear 1 with gear 2 was a single ellipse and the gear 1 meshed with gear 2 was a stability state in this situation. The cycle time was doubled in this gear transmission state as the Poincare map had two red points. The gear 1 meshed with gear 3 had the same cyclic motion state, but the vibration amplitude of gear 1 with gear 3 was larger than the gear 1 with gear 2.

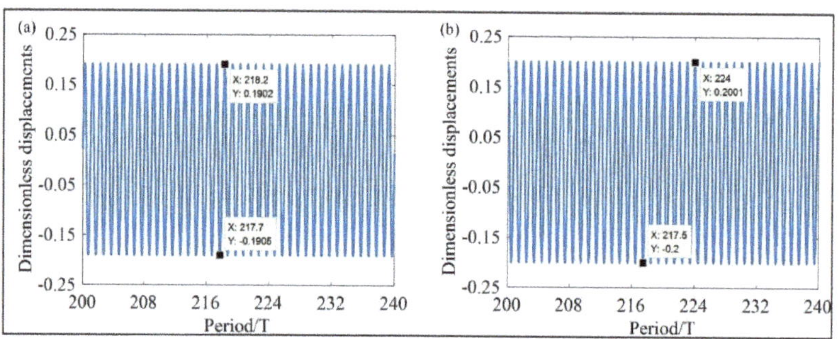

Figure 3. (**a**) The dimensionless relative displacement of the gear 1 and gear2, (**b**) the dimensionless relative displacement of the gear 1 and gear 3.

In order to investigate the effect of the angle on the adherence force, the theoretical model of this claw when climbing the rough surface is shown in Figure 5d. In this figure, two small circles represent the two tips of the spine, which contacts the rough surface. The radius of the two smaller circles is r. The bigger circle is the particle on the rough surface, whose radius is R. These two small circles are tangent to the big circle. They express the cicada's claw adherence to the rough surface. The angle between these two circles is γ, which represents the angle of the hooks on the end of the leg. The contract angle between the cicada's claw and the rough surface is α_i (α_1, α_2).

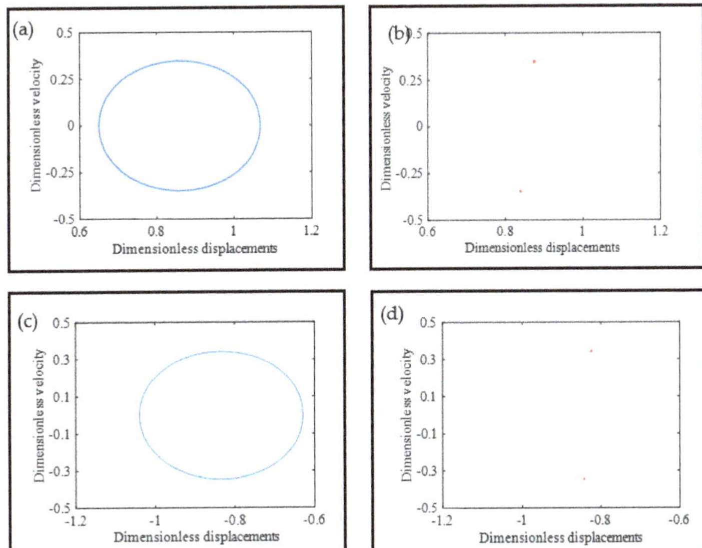

Figure 4. (**a**) The phase images of the gear 1 and gear 2. (**b**) The Poincare maps of the gear 1 and gear 2. (**c**) The phase images of the gear 1 and gear 3. (**d**) The Poincare maps of the gear 1 and gear 3.

2.4. Adhesion Part Analysis

Figure 5a is the cicada, which has the ability to climb trees by using its biometry claws. Figure 5b represents the detailed image of the cicada's claw. From this image, we can see a special bifurcated spine on the end of the claw. The cicada can climb trees by using these bifurcated spines. This paper investigated the use of bifurcated spines in a proposed wall-climbing robot for climbing on rough surfaces. This paper presents a bionic spine design that was inspired by this special hook with an angle, as shown in Figure 5c.

The contact angle α is

$$\alpha = arcsin(\frac{r + R - h}{r + R}) = arcsin(1 - \frac{h}{r + R}) \tag{29}$$

Figure 5e expresses the equivalent force diagram of the bionic hook adhering to the rough surface. The load distribution force angle is β (β_1, β_2), the height of the particle is h, f (f_1, f_2) is the friction force between hook and surface, F (F_1, F_2) is the comprehensive capsizing force of the hook by the wall and the weight of the hook, and N (N_1, N_2) is the support force on the hook by the wall. To adhere to the protruding particles, the mechanics must satisfy:

$$\begin{cases} \sum_{i=1}^{n} f_i sin(\gamma_i/2) = \sum_{i=1}^{n} \mu_i N_i cos(\gamma_i/2) \\ \qquad = \sum_{i=1}^{n} F_i sin(\alpha_i + \beta_i) \\ \\ \sum_{i=1}^{n} N_i = \sum_{i=1}^{n} F_i cos(\alpha_i + \beta_i) \end{cases} \tag{30}$$

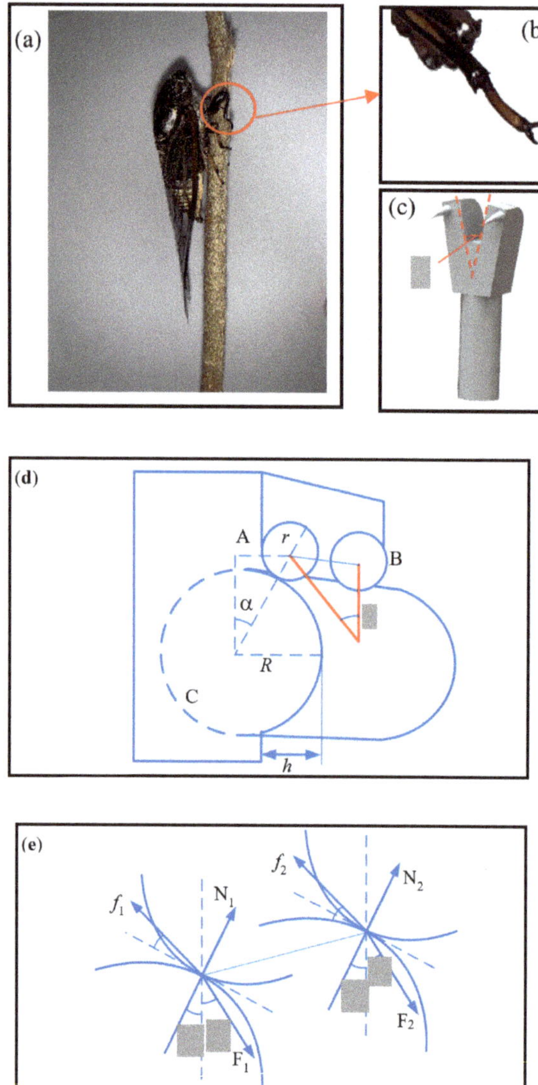

Figure 5. (**a**) A cicada on the branch of the tree, (**b**) a leg of a cicada, (**c**) the bionic bifurcated spines inspired by the cicada's claws, (**d**) equivalent force diagram of the bionic hook. (**e**) The equivalent force diagram of the bionic hook adhering to the rough surface.

In terms of a single hook,

$$\beta < arctan(2\mu cos(\gamma/2)) - \alpha \tag{31}$$

Here, μ is the friction coefficient between the hook and the surface. Integrating with Equations (29)–(31), β is written as

$$\beta < arctan(2\mu cos(\gamma/2)) - arcsin(1 - \frac{h}{r + R}) \tag{32}$$

The hooks' adherence on the rough surface must satisfy the condition that β is greater than zero. This means that α is limited and is less than $arctan(\mu cos(\gamma/2))$, from Equation (31). Here, α is defined by the diameter and the height of the particle and α is the angle of the hook to change the adhere characteristics of the hooks. These parameters are investigated by the following figures. Figure 6a represents the relations of the pressure angle (α) and the load distribution angle (β) with various hooks' angles γ. In this figure, the load distribution angle (β) decreases as the hook angles γ decrease. It means that decreasing the hook angles γ improves the adherence property. The relation between the size of hooks and the particles on the contact surface is provided in Figure 6b. In this figure, the smaller the radius of the hooks' angle, the bigger the load distribution angle (β), as the height and the radius of the particles remain unchanged. Hence, decreasing the size of the hook's end can develop the adherence property.

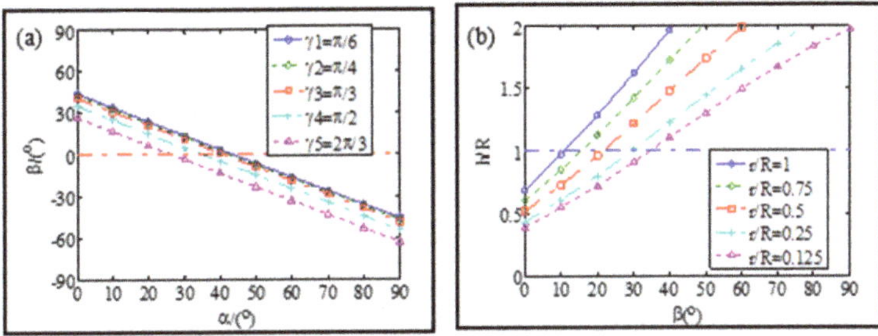

Figure 6. (**a**) The relation of the pressure angle (α) and the load distribution angle (β) with various hook angles (γ), (**b**) the relation of the load distribution load angle and the particles' sizes on the contact surface with different sizes of hooks.

2.5. Adhesion Part Analysis

As shown in Figure 7, the tangential adhesive force of the various hook angles of the bionic claws were examined on the adhesive test platform. A rough surface was put on the platform for testing.

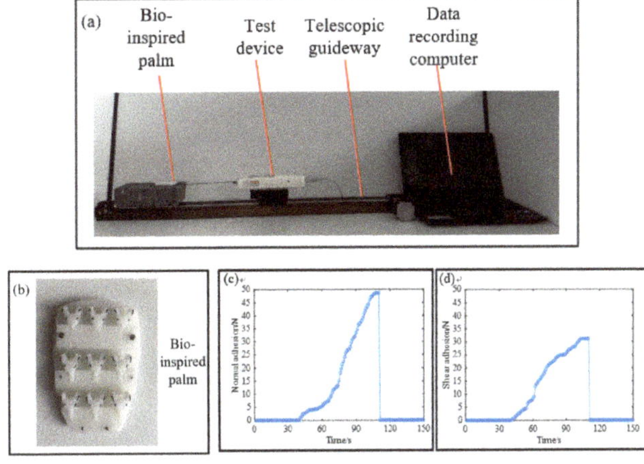

Figure 7. The adhesion force of the bionic palm. (**a**) A measuring platform to measure the adhere force. (**b**) The bionic palm with various bionic spines. (**c**) The normal adhesion force of this bionic palm. (**d**) The shear adhesion force of this bionic palm.

Figure 7a represents the testing set for adhesive materials. The testing device and the telescopic guideway are on the right; the testing materials are on the left. The adherence palm that was bio-inspired by the cicada (the hook angle is 60°) was put on the brick that was fixed on the platform. The testing device was a force test set that could record the data during the testing. The testing device was fixed on the telescopic guideway with a motor pulling the bionic palm and recording the force data. The preload of the robot's palm was 0.12 N on the wall during climbing. Hence, the preload on the adherence palm was also 0.12 N during the measuring. Figure 7c,d shows the normal adhesion force and the shear adhesion force of the bionic palm (Figure 7b), respectively. The maximum of the normal adhesion force reached 48.7 N, and the maximum of the shear adhesion force reached 31.5 N.

To adhere to the smooth surface, we borrowed the adhesion characteristics of a gecko's toes with numerous bristles' arrays. Figure 8a is a gecko from an animal breeding base. Figure 8b is the gecko's toe and the Figure 8c is the scanning electron microscope image of the gecko's toe. From Figure 8c, there are numerous bristles' arrays on the toe to help the gecko adhere to the wall for climbing.

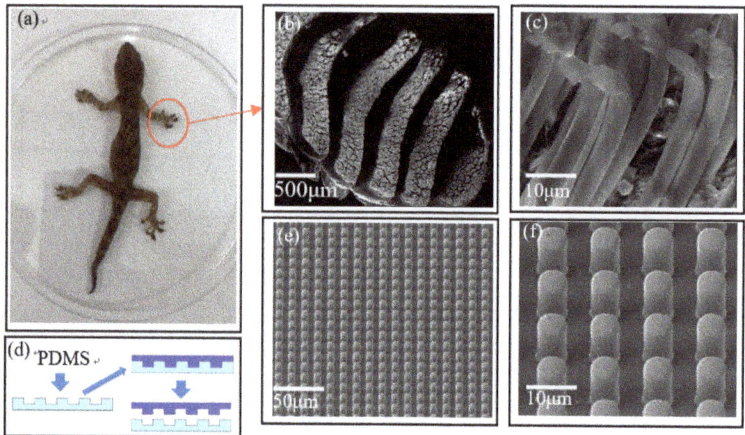

Figure 8. (**a**) The photo of the gecko, (**b**) scanning electron micrograph of a gecko's toe, (**c**) the extended scanning electron micrograph of a gecko's toe. (**d**) The PDMS was taken from the silicon plate to obtain an adhesion sheet. (**e**) The gecko-inspired setae array, (**f**) the extended image of the gecko-inspired setae array.

Furthermore, the photolithography technology was used to make the bionic bristles' array (Figure 8d). This bionic bristles' array is made of PDMS (Polydimethylsiloxane), as shown in Figure 8e,f. The diameter of the bristle was 7 μm and the height was 10 μm. After the adhesive sheet was manufactured, a 40-mm × 50-mm rectangle sheet was cut for measuring. The result is presented in Figure 9.

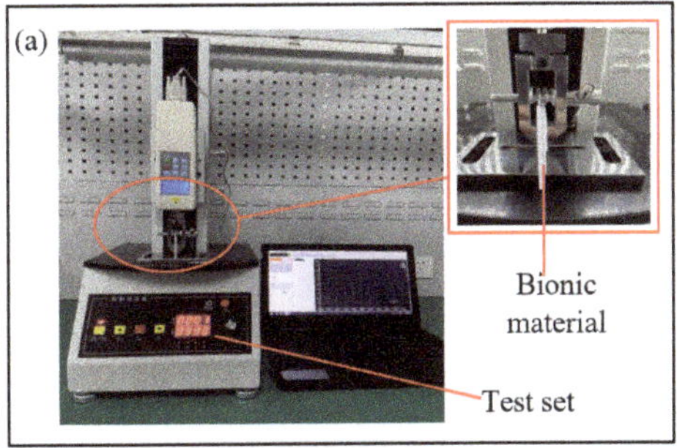

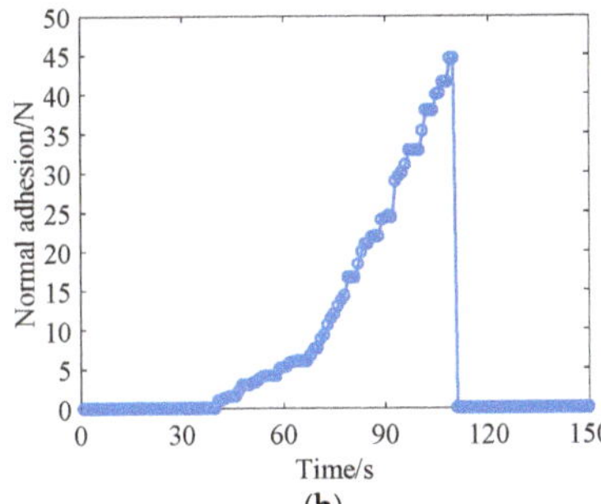

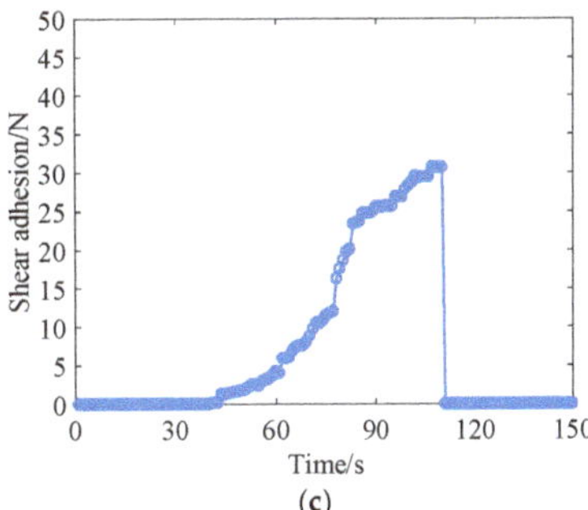

Figure 9. The test set to measure the adherence force. (**a**) The test set for measuring the adherence force of the bio-inspired materials. (**b**) The test set to measure the normal force of the adhesive materials. (**c**) The shearing force of the adhesive materials without preload.

The adherence materials (40 mm × 50 mm) that were put on the glass fixed on the platform, as shown in Figure 9a–c, showed normal adherence force and shear adherence force, respectively. The maximum of the normal adherence materials reached 44.5 N, and the maximum of the shear adherence force reached 30.7 N. This rectangle sheet was manufactured for the bionic robot to adhere to the smooth surface.

3. Locomotion of the Robot

The proposed robot climbed on the wall with stability by an efficient locomotion gait, as shown in Figure 10. This climbing strategy was divided into six parts by the servos to realize the robot's legs lifting and stretching. In each period, the foot (I) and the foot (IV) moved in sync, and the foot (II) and foot (III) moved in sync. The foot with blue color means that the foot was in the attaching status, while the white color indicates it was in the detaching status. However, each foot was in two positions when the robot climbed the wall: One was at the stretching position and the other was at the contraction position. The stretching position was the point of the robot's foot after the stretching was completed and the contraction position was the point of the robot's foot after contracting was completed.

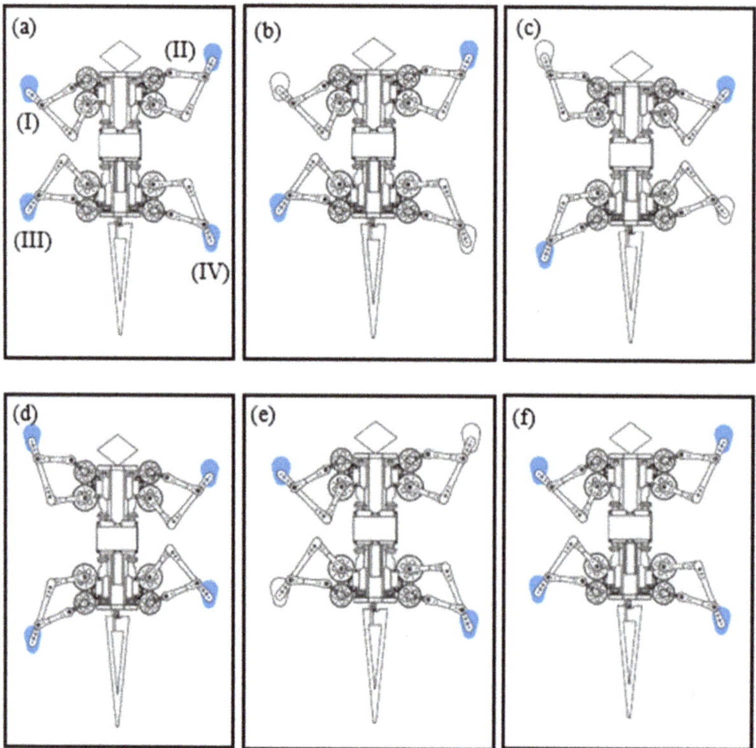

Figure 10. (**a**) The first step in a single cycle, (**b**) the second step in a single cycle, (**c**) the third step in a single cycle, (**d**) the fourth step in a single cycle, (**e**) the fifth step in a single cycle, and (**f**) the sixth step in a single cycle.

In the first step (Figure 10a), the four feet of the robot all attached to the surface. The foot (I) and the foot (IV) were at the contraction position and the foot (II) and the foot (III) were at the stretching position.

In the second step (Figure 10b), the foot (II) and the foot (III) remained attached at the above position. Then, the foot (I) and the foot (IV) lifted away from the surface by the servos.

In the third step (Figure 10c), the foot (II) and the foot (III) were attached at the above position and the servo drove the gear transmission system to elevate the body. Meanwhile, the servo drove the gear transmission system to stretch the foot (I) and the foot (IV).

In the fourth step (Figure 10d), the foot (II) and the foot (III) were attached at the above position. The servo drove the foot (I) and the foot (IV) down to attach to the surface. The four feet were all attached on the surface in this step.

In the fifth step (Figure 10e), the servo drove the gear transmission to the control foot (I) and the foot (IV) to the contraction position for elevating the body. At the same time, the servo uplifted the foot (II) and the foot (III) and detached from the surface to the stretching position.

In the last step (Figure 10f), the foot (I) and the foot (IV) were kept attached at the above contraction position. The servo drove the foot (II) and the foot (III) down and attached them to the surface.

4. Experiment

The dimensions of the gear transmission robot prototype are shown in Table 2. The components of this bionic, wall-climbing robot were made from resin by a 3D printing method. The manufactured parts were assembled with the adhesion hooks (Figure 7b) and the adhesion materials (Figure 8e) to complete the climbing experiment as follow.

Table 2. The dimensions of the robot prototype.

Design Parameters	Values
Length of robot (with tail)	474 mm
Width of body (include the legs)	330 mm
Mass of robot	360 g
Size of pad	40 mm × 50 mm
One-step duration	0.15 s

The locomotion of this robot climbing the vertical cloth, stones, and glass walls in a cycle is shown in Figures 11–13, respectively. This robot climbed from the dotted line to the real line. The laser position device recorded the displacement of the robot in a cycle, as shown in Figure 14. The robot climbing 90.10 mm on the horizontal cloth surface and 79.7mm on the vertical cloth surface is presented in Figure 14b (i) and (ii). The locomotion time of each step was 0.15 s. Then, the six steps took 0.9 s in a cycle. The theoretical speed of the robot was 11.34 cm/s, and the experimental were 10.01 cm/s and 8.86 cm/s on the horizontal and vertical wall, respectively. The theoretical speed of the robot when climbing walls was higher than the experimental results. The difference between theory and experimental was due to the robot assembling error. The displacements of the robot in a cycle on the stones and glass are shown in Figure 14b (iii) and (iv), respectively. The robot climbed 82.29 mm on the vertical stone surface and 84.02 mm on the vertical glass surface. The climbing speeds of the robot on the stone wall and glass wall were 9.14 cm/s and 9.34 cm/s, as shown in Table 3, respectively. When the wall-climbing robot climbed the cloth surface, the cloth surface was slightly damaged, and when it climbed the stone surface, the hook stabbing was slightly damaged. The different climbing speeds of the robot may have been due to the surface materials. The robot can adapt to different surfaces by changing the adhesion materials. Research on multiple series of materials can broaden the scope of application of robots.

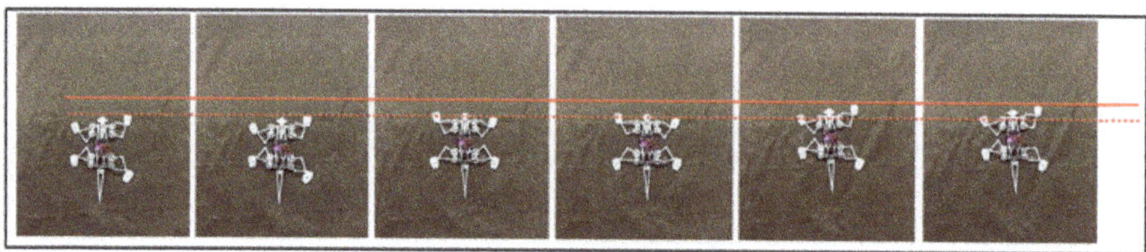

Figure 11. Experiment of the locomotion of the bionic robot climbing on the cloth surface in cycle.

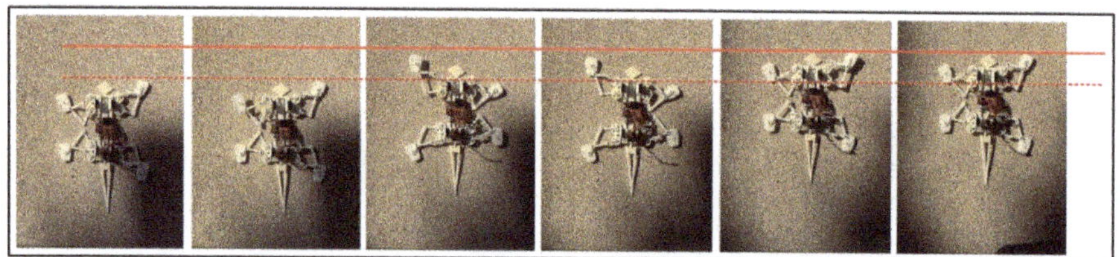

Figure 12. Experiment of the locomotion of the bionic robot climbing on the stone surface in cycle.

Figure 13. Experiment of the locomotion of the bionic robot climbing on the glass surface in cycle.

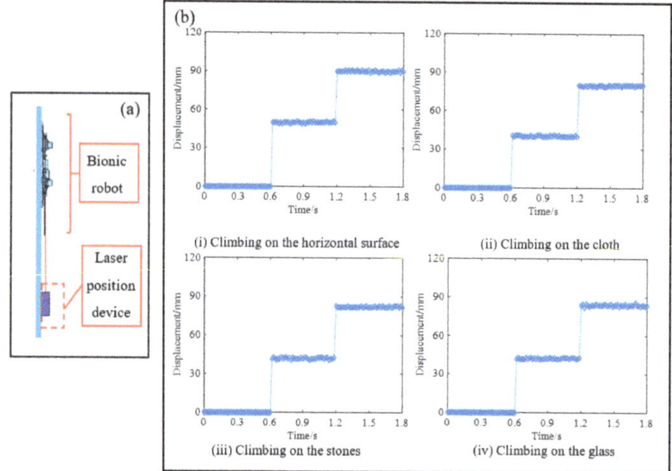

Figure 14. (**a**) Experiment of the locomotion speed of the bionic robot by the laser position device, (**b**) the displacement profile of the robot.

Table 3. The speed of the robot prototype.

Climbing Surfaces	Velocity (Vertical)
Cloth	8.86 cm/s
Stones	9.14 cm/s
Glass	9.34 cm/s

Table 4 lists the speed parameters of various wall-climbing robot. Compared with these climbing robots (Waalbot II, Stickbot, Spinybot II, as shown in Table 4), it is rarely adapted to multi-surfaces. Waalbot II and Stickbot were designed for smooth surfaces, while Spinybot II was designed for rough surfaces. Few robots are designed to climb multiple walls. This paper's robot climbed on the smooth and rough surfaces. The speed of this robot was 8.86 cm/s, 9.14 cm/s, and 9.34 cm/s on the cloth, stones, and glass. Compared to these climbing robots in Table 4, this paper's robot expands the scope of application and performs well in terms of climbing speed.

Table 4. The comparison of the robots.

Wall-Climbing Robot	Climbing Surfaces	Velocity (Vertical)	Climbing on Multi-Surfaces
Waalbot II [10]	Glass	5 cm/s	No
Stickbot [11]	Glass	4 cm/s	No
Spinybot II [19]	Stones	4 cm/s	No

5. Conclusions

In this paper, a wall-climbing robot that can adhere to rough and smooth surfaces inspired by cicadas and geckos was proposed. A new type of climbing structure was designed with a gear-link transmission to improve the stability of the robot when climbing a wall. This robot had a special angle spine adhesion palm for the rough surface that was inspired by the cicada. Meanwhile, the robot palm used adhesion materials, inspired by the gecko, for smooth surfaces. This palm design allowed the robot to effectively adhere to a wall. The robot was experimentally evaluated on vertical rough and smooth surfaces. It reached a speed of 8.86 cm/s on the cloth surface, 9.14 cm/s on the stone surface, and 9.34 cm/s on the glass surface. Compared to other robots, this robot had a good performance. This type of robot provided an idea for the research of wall-climbing robots that deal with a variety of wall surfaces. This robot has practical applications, including inspecting rough and smooth surfaces.

Author Contributions: Conceptualization, D.K.; writing—original draft preparation, S.B.; validation, F.X.; formal analysis, Y.W.; All authors have read and agreed to the published version of the manuscript.

Funding: This research received no external funding.

Institutional Review Board Statement: Ethical review and approval were waived for this study, due to this paper don't involve ethical issue.

Informed Consent Statement: Not applicable.

Data Availability Statement: The study did not involve important data.

Conflicts of Interest: The authors declare no conflict of interest.

References

1. Maderson, P.F. Keratinized epidermal derivatives as an aid to climbing in gekkonid lizards. *Nature* **1964**, *203*, 780–781. [CrossRef]
2. Hiller, U. Untersuchungen zum Feinbau und zur Funktion der Haftborsten von Reptilien. *Z. Morphol. Tiere* **1968**, *62*, 307–362. [CrossRef]

3. Russell, A.P. A contribution to the functional morphology of the foot of the tokay, Gekko gecko (Reptilia, Gekkonidae). *J. Zool.* **2009**, *176*, 437–476. [CrossRef]
4. Autumn, K.; Liang, Y.A.; Hsieh, S.T.; Zesch, W.; Chan, W.P.; Kenny, T.W.; Fearing, R.; Full, R.J. Adhesive force of a single gecko foot-hair. *Nature* **2000**, *405*, 681–685. [CrossRef] [PubMed]
5. Autumn, K.; Sitti, M.; Liang, Y.A.; Peattie, A.M.; Hansen, W.R.; Sponberg, S.; Kenny, T.W.; Fearing, R.; Israelachvili, J.N.; Full, R.J. Evidence for van der Waals adhesion in gecko setae. *Proc. Natl. Acad. Sci. USA* **2002**, *99*, 12252–12256. [CrossRef] [PubMed]
6. Unver, O.; Uneri, A.; Aydemir, A.; Sitti, M. Geckobot: A gecko inspired climbing robot using elastomer adhesives. In Proceedings of the IEEE International Conference on Robotics and Automation, Orlando, FL, USA, 15–19 May 2006; pp. 2329–2335.
7. Carlo, M.; Metin, S. A Biomimetic Climbing Robot Based on the Gecko. *J. Bionics Eng.* **2006**, *3*, 115–125.
8. Murphy, M.P.; Aksak, B.; Sitti, M. Gecko-Inspired Directional and Controllable Adhesion. *Small* **2009**, *5*, 170–175. [CrossRef] [PubMed]
9. Aksak, B.; Murphy, M.P.; Sitti, M. Gecko Inspired Micro-Fibrillar Adhesives for Wall-climbing Robots on Micro/Nanoscale Rough Surfaces. In Proceedings of the IEEE International Conference on Robotics & Automation, Pasadena, CA, USA, 19–23 May 2008; pp. 3058–3063.
10. Murphy, M.P.; Kute, C.; Menguc, Y.; Sitti, M. Waalbot II: Adhesion Recovery and Improved Performance of a Climbing Robot, using Fibrillar Adhesives. *Int. J. Robot. Res.* **2011**, *30*, 118–133. [CrossRef]
11. Kim, S.; Spenko, M.; Trujillo, S.; Heyneman, B.; Santos, D.; Cutkosky, M.R. Smooth vertical surface climbing with directional adhesion. *IEEE Trans. Robot.* **2008**, *1*, 65–74.
12. Santos, D.; Heyneman, B.; Kim, S.; Esparza, N.; Cutkosky, M.R. Gecko-inspired climbing behaviors on vertical and overhanging surfaces. In Proceedings of the International Conference on Robotics & Automation, IEEE, Pasadena, CA, USA, 19–23 May 2008; p. 1125.
13. Asbeck, A.; Dastoor, S.; Parness, A.; Fullerton, L.; Esparza, N.; Soto, D.; Heyneman, B.; Cutkosky, M.R. Climbing rough vertical surfaces with hierarchical directional adhesion. In Proceedings of the IEEE International Conference on Robotics and Automation, Kobe, Japan, 12–17 May 2009; pp. 2675–2681.
14. Hawkes, E.W.; Ulmen, J.; Esparza, N.; Esparza, N.; Cutkosky, M.R. Scaling walls: Applying dry adhesives to the real world. In Proceedings of the International Conference on Intelligent Robots and Systems, San Francisco, CA, USA, 25–30 September 2011; pp. 5100–5106.
15. Yu, Z.W.; Wang, Z.Y.; Liu, R.; Wang, P.; Dai, Z.D. Stable gait planning for a gecko-inspired robot to climb on vertical surface. In Proceedings of the International Conference on Mechatronics and Automation, Takamatsu, Japan, 4–7 August 2013; pp. 307–311.
16. Wu, X.; Wang, D.P.; Zhao, A.W.; Li, D.; Mei, T. A Wall-Climbing Robot with Biomimetic Adhesive Pedrail. In *Advanced Mechatronics and MEMS Devices*; Springer: New York, NY, USA, 2013; Volume 23, pp. 179–191.
17. Zhang, Y.J.; Wu, Y.X.; Liu, Y.W.; Hu, C.Y.; Sun, S.M.; Mei, T. Bionic Design of the Body of Tank-like Climbing Robot. In Proceedings of the International Conference on Mechatronics and Automation, Chengdu, China, 5–8 August 2012; pp. 2287–2291.
18. Dai, Z.D.; Gorb, S.N.; Schwarz, U. Roughness-dependent friction force of the tarsal claw system in the beetle Pachnoda marginata. *J. Exp. Biol.* **2002**, *205*, 2479–2488. [CrossRef] [PubMed]
19. Kim, S.; Asbeck, A.T.; Cutkosky, M.R.; Provancher, W.R. SpinybotII: Climbing Hard Walls with Compliant Microspines. In Proceedings of the 12th International Conference on, Seattle, WA, USA, 18–20 July 2005; pp. 601–606.
20. Asbeck, A.T.; Kim, S.; Cutkosky, M.R.; Provancher, W.; Lanzetta, M. Scaling Hard Vertical Surfaces with Compliant Microspine Arrays. *Int. J. Robot. Res.* **2006**, *25*, 1165–1179. [CrossRef]
21. Alan, A.T.; Kim, S.; McClung, A.; Parness, A.; Cutkosky, M.R. Climbing walls with microspines. In Proceedings of the International Conference on Robotics & Automation, Florida, IL, USA, 15–19 January 2006; pp. 4315–4317.
22. Saunders, A.; Goldman, D.; Full, R.J.; Buehler, M. The rise climbing robot: Body and leg design. *Def. Secur. Symp.* **2006**, *6230*, 1–13.
23. Spenko, M.J.; Haynes, G.C.; Saunders, J.A.; Cutkosky, M.R.; Rizzi, A.; Rizzi, A.A.; Full, R.J.; Koditschek, D.E. Biologically inspired climbing with a hexapedal robot. *J. Field Robot.* **2008**, *25*, 223–242. [CrossRef]
24. Haynes, G.C.; Khripin, A.; Lynch, G.; Amory, J.; Saunders, A.; Rizzi, A.A.; Koditschek, D.E. Rapid Pole Climbing with a Quadrupedal Robot. In Proceedings of the International Conference on Robotics and Automation, Kobe, Japan, 12–17 May 2009; pp. 12–17.
25. Goldman, D.I. Dynamics of rapid vertical climbing in cockroaches reveals a template. *J. Exp. Biol.* **2006**, *15*, 2990–3000. [CrossRef] [PubMed]
26. Clark, J.E.; Goldman, D.I.; Lin, P.C.; Lynch, G.; Chen, T.S.; Komsuoglu, H.; Full, R.J.; Koditschek, D. Design of a Bio-inspired Dynamical Vertical Climbing Robot. In Proceedings of the International Conference on Robotics: Science and Systems, Atlanta, GA, USA, 27–30 June 2007; pp. 9–16.
27. Lynch, G.A.; Clark, J.E.; Koditschek, D. A self-exciting controller for high-speed vertical running. In Proceedings of the 2009 IEEE/RSJ International Conference on Intelligent Robots and Systems, St. Louis, MO, USA, 10–15 October 2009; pp. 631–638.
28. Dickson, J.D.; Clark, J.E. The effect of sprawl angle and wall inclination on a bipedal, dynamic climbing platform. In Proceedings of the Fifteenth International Conference on Climbing and Walking Robots and the Support Technologies for Mobile Machines, Baltimore, MD, USA, 23–26 July 2012; pp. 459–466.
29. Lynch, G.A.; Clark, J.E.; Lin, P.C.; Koditschek, D.E. A bioinspired dynamical vertical climbing robot. *Int. J. Robot. Res.* **2012**, *31*, 974–996. [CrossRef]

30. Dickson, J.D.; Patel, J.; Clark, J.E. Towards maneuverability in plane with a dynamic climbing platform. In Proceedings of the International Conference on Robotics and Automation, Karlsruhe, Germany, 8–13 May 2013; pp. 1355–1361.
31. Miller, B.; Ordonez, C.; Clark, J.E. Examining the effect of rear leg specialization on dynamic climbing with SCARAB: A dynamic quadrupedal robot for locomotion on vertical and horizontal surfaces. *Exp. Robot.* **2013**, *88*, 113–126.
32. Miller, B.; Clark, J.; Darnell, A. Running in the horizontal plane with a multi-modal dynamical robot. In Proceedings of the International Conference on Robotics and Automation, Karlsruhe, Germany, 6–10 May 2013; pp. 3335–3341.
33. Wile, G.D.; Daltorio, K.A.; Diller, E.D.; Palmer, L.R.; Gorb, S.N.; Ritzmann, R.E.; Quinn, R.D. Screenbot: Walking inverted using distributed inward gripping. In Proceedings of the International Conference on Intelligent Robots and Systems, Nice, France, 22–26 September 2008; pp. 1513–1518.
34. Liu, Y.; Sun, S.; Wu, X.; Mei, T. A Wheeled Wall-Climbing Robot with Bio-Inspired Spine Mechanisms. *J. Bionic Eng.* **2015**, *12*, 17–28. [CrossRef]
35. Liu, Y.W.; Liu, S.W.; Mei, T.; Wu, X.; Li, Y. Design and Analysis of a Bio-Inspired Tracked Wall-Climbing Robot with Spines. *Robot* **2019**, *41*, 527–533.

applied sciences

Article

Control Strategy for Direct Teaching of Non-Mechanical Remote Center Motion of Surgical Assistant Robot with Force/Torque Sensor

Minhyo Kim, Youqiang Zhang and Sangrok Jin *

School of Mechanical Engineering, Pusan National University, Pusan 46241, Korea; mhkim1@pusan.ac.kr (M.K.); zhangyq@pusan.ac.kr (Y.Z.)
* Correspondence: rokjin17@pusan.ac.kr; Tel.: +82-51-510-2984

Featured Application: Surgical robot.

Abstract: This paper presents a control strategy that secures both precision and manipulation sensitivity of remote center motion with direct teaching for a surgical assistant robot. Remote center motion is an essential function of conventional laparoscopic surgery, and the most intuitive way a surgeon manipulates a robot is through direct teaching. The surgical assistant robot must maintain the position of the insertion port in three-dimensional space during the four-degree-of-freedom motions such as pan, tilt, spin, and forward/backward. In addition, the robot should move smoothly when controlling it with the hands during the surgery. In this study, a six-degree-of-freedom collaborative robot performs the cone-shaped trajectory with pan and tilt motion of an end-effector keeping the position of the remote center. Instead of the bulky mechanically constrained remote center motion mechanism, a conventional collaborative robot is used to mimic the wrist movement of a scrub nurse. A force/torque sensor that is attached between the robot and end-effector estimates the surgeon's intention. A direct teaching control strategy based on position control is applied to guarantee precise remote center position maintenance performance. A motion generation algorithm is designed to generate motion by utilizing a force/torque sensor value. The parameters of the motion generation algorithm are optimized so that the robot can be operated with uniform sensitivity in all directions. The precision of remote center motion and the torque required for direct teaching are analyzed through pan and tilt motion experiments.

Keywords: surgical assistant robot; remote center motion; direct teaching; impedance control

1. Introduction

Robotic surgery is the latest trend in minimally invasive surgery. The Da Vinci System (Intuitive Surgical Inc., Sunnyvale, CA, USA) is well known as the world's most successful surgical robot. As the Da Vinci System is a remote-control console in which the operator controls the robot, the operator has no choice but to separate it from the operating table. However, many surgeons want to be able to freely intervene and flexibly cope with unexpected situations during robotic surgery. Since surgical robots are super expensive, only a small number of tertiary hospitals have them, and they are used for serious surgery considering the overall procedures such as installation and recovery of robots during surgery. Therefore, considering the entire medical field including the primary hospital, a robot that can be used for laparoscopic surgery, which requires two hours of operation, is needed. In addition, medical personnel are complaining of fatigue from surgery due to a shortage of medical assistants, and demands for surgical assistance robots as shown in Figure 1a are increasing.

Citation: Kim, M.; Zhang, Y.; Jin, S. Control Strategy for Direct Teaching of Non-Mechanical Remote Center Motion of Surgical Assistant Robot with Force/Torque Sensor. *Appl. Sci.* 2021, 11, 4279. https://doi.org/10.3390/app11094279

Academic Editor: Manuel Armada

Received: 15 April 2021
Accepted: 6 May 2021
Published: 9 May 2021

Publisher's Note: MDPI stays neutral with regard to jurisdictional claims in published maps and institutional affiliations.

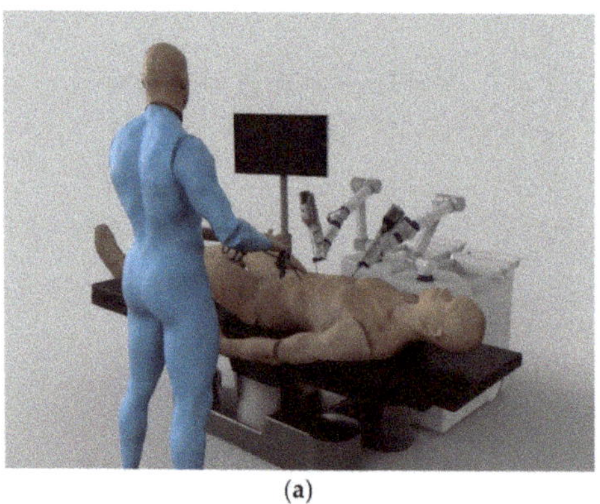 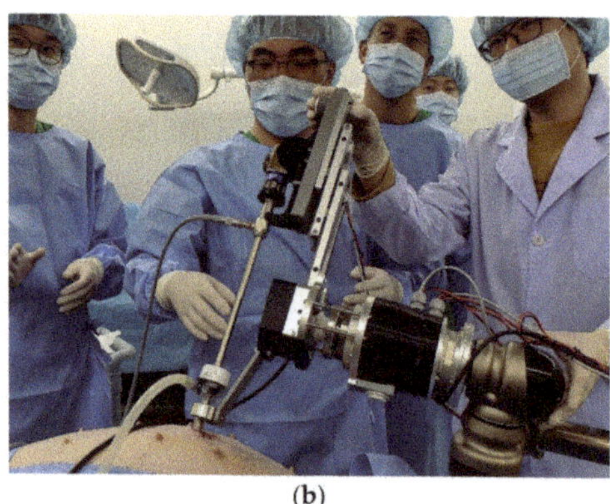

(a)   (b)

Figure 1. Introduction of the surgical assistant robot: (**a**) conceptual figure; (**b**) animal experiment.

In this study, we developed an entry-level surgical assistant robot with an end-effector attached to a six-joint cooperative robot [1]. The robot works with the surgeon on the operating table by holding a laparoscopic camera and pulling the internal tissue with forceps. To evaluate the validity and usability of the design of the prototype, animal experiments were conducted using pigs as shown in Figure 1b. The most basic function of laparoscopic surgery assistance robots is remote center motion (RCM), and direct teaching is provided on how the operator can intuitively manipulate them without additional training. The stability of RCM and the sensitivity to direct teaching are important indicators for performance evaluation from the point of view of surgeons.

The movement of an imaginary point outside the organization is called remote center motion. The RCM is an essential function for maintaining the position of the insertion port while the surgical robot inserts the surgical tool into the human body with a trocar and operates it. Medical robots have a variety of mechanisms for RCM [2]. The method of mechanically limiting the remote center and always maintaining the remote center in any movement of the robot is widely used for high safety. Many surgical robots, including Da Vinci, maintain a remote center by decoupled motion in a z-bar-shaped parallelogram link structure [3,4]. Some robots maintain the remote center by using circular guides [5,6]. Microsurgical robots such as eye surgery have parallel mechanisms with closed linkage [7]. However, the mechanically constrained system is not suitable for a surgical assistant robot that works in the same space as the operator because it consists of parallel mechanisms and support mechanisms and has a large possibility of colliding with the surgeon when placed on the operating table. To minimize the size of the robot, there is also a non-mechanical method of maintaining a virtual remote center through the control of articulated robots [8].

With the recent development of cooperative robots, various direct teaching methods are being studied. The most widely used manual traction method for low-cost cooperative robots is to control current without sensors [9,10]. However, since it operates only when the applied torque is large enough considering the reduction ratio and friction of the reduction gear, it does not meet the requirements of the surgical site, and because it is not position-based control, remote center maintenance performance is unreliable. There is a system that allows sensitive and precise operation by attaching a torque sensor to each rotating shaft of the robot, but it is mainly used for expensive cooperative robots [11,12]. Another method is to attach a force/torque sensor to the tip of a cooperative robot and teach it directly based on the sensor [13,14]. The robot has the disadvantage of being able to teach softly by taking the tip of the robot, but not by using other links in the middle.

In this study, a six-degree-of-freedom force/torque sensor is attached to the tip of the cooperative robot to connect the end-effector. A method of maintaining a non-mechanical remote center through control, measuring the force and torque applied by the user by grasping the end-effector, and directly using it for instruction is used. The value generated by the force/torque sensor is converted into a control command and input into position control, so the remote center can be kept solid, but care must be taken in the design of the motion generation algorithm to create a sensitive feeling. Through an experiment in which the user directly teaches by grasping the end-effector of the robot, it evaluates the ability to follow instructions well and the performance to which it maintains the remote center position.

2. Non-mechanical Remote Center Motion

2.1. System Configuration for Remote Center Motion

In general laparoscopic surgery, the surgical instrument requires four-degrees-of-freedom motion: Pan (Roll), Tilt (Pitch), Spin (Yaw) rotational and linear movements (Forward/Backward), as shown in Figure 2a. In this study, a two-degree-of-freedom end-effector is mounted on a six-degree-of-freedom cooperative robot, and a virtual remote center is defined by a control algorithm to achieve RCM. A six-degree-of-freedom cooperative robot embodies Pan and Tilt rotational movements in the four-degree-of-freedom RCM while maintaining the remote center of three-dimensional space to form a conical trajectory. Surgery assistance is performed in such a small space that the scrub nurse has to hold an endoscopic camera between the sides of a surgeon. This robot, which maintains a non-mechanical remote center, mimics the movement of a human wrist, as shown in Figure 2a. In this case, a small device mechanism can have the effect of human grasping surgical tools in a narrow space and minimize collisions when collaborating with surgeons. However, the robustness of the controller is very important for safe surgery. As the two-degree-of-freedom motion of an end-effector can be controlled independently, this paper focuses research on the pan, tilt RCM control of the cooperative robot.

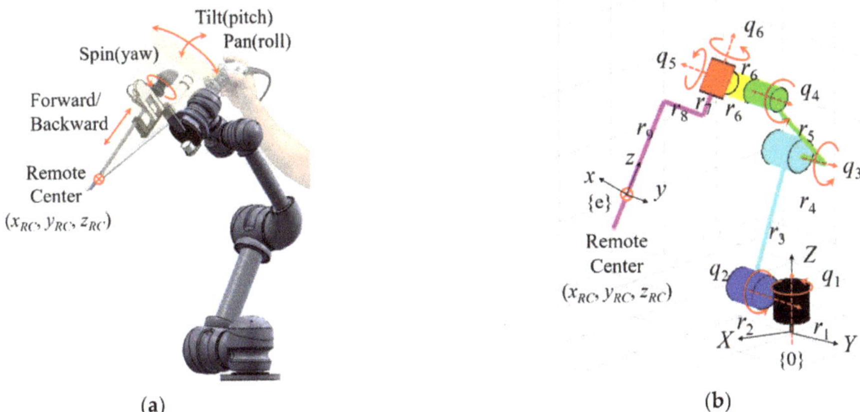

Figure 2. Kinematic model of the surgical assistant robot: (**a**) remote center motion; (**b**) configuration of joints and links.

2.2. Kinematic Modeling

The kinematic model of a robot is required as shown in Figure 2b. A robot base has a fixed frame {0} and a moving frame {e} is attached to a remote center point, and the rotation angle of each joint is q_i and the link parameter is r_i. The kinematic model of a robot is derived using a homogeneous transformation matrix between moving frames attached to each joint as shown in Equation (1). The forward kinematics of a robot manipulator can be obtained by a sequential product of transformation matrices as shown in Equation (2). Coordinates of the remote center point in the three-dimensional space, (x_{RC}, y_{RC}, z_{RC}), can be extracted from the fourth column of the matrix $\mathbf{T}_{0e}$ as shown in Equation (3), and the pan and tilt angles of the end-effector can be calculated from the rotation matrix elements of the matrix $\mathbf{T}_{0e}$ as shown in Equation (4). The link parameters of the robot manipulator used in this study are shown in Table 1.

$$
\mathbf{T}_{01} = \begin{bmatrix} \cos q_1 & -\sin q_1 & 0 & 0 \\ \sin q_1 & \cos q_1 & 0 & 0 \\ 0 & 0 & 1 & r_1 \\ 0 & 0 & 0 & 1 \end{bmatrix}, \quad
\mathbf{T}_{12} = \begin{bmatrix} \cos q_2 & 0 & \sin q_2 & 0 \\ 0 & 1 & 0 & -r_2 \\ -\sin q_2 & 0 & \cos q_2 & 0 \\ 0 & 0 & 0 & 1 \end{bmatrix},
$$

$$
\mathbf{T}_{23} = \begin{bmatrix} \cos q_3 & 0 & \sin q_3 & 0 \\ 0 & 1 & 0 & 0 \\ -\sin q_3 & 0 & \cos q_3 & r_3 \\ 0 & 0 & 0 & 1 \end{bmatrix}, \quad
\mathbf{T}_{34} = \begin{bmatrix} \cos q_4 & 0 & \sin q_4 & 0 \\ 0 & 1 & 0 & r_4 \\ -\sin q_4 & 0 & \cos q_4 & r_5 \\ 0 & 0 & 0 & 1 \end{bmatrix},
$$

$$
\mathbf{T}_{45} = \begin{bmatrix} 1 & 0 & 0 & 0 \\ 0 & \cos q_5 & -\sin q_5 & -r_6 \\ 0 & \sin q_5 & \cos q_5 & 0 \\ 0 & 0 & 0 & 1 \end{bmatrix}, \quad
\mathbf{T}_{56} = \begin{bmatrix} \cos q_6 & -\sin q_6 & 0 & -r_6 \\ \sin q_6 & \cos q_6 & 0 & 0 \\ 0 & 0 & 1 & 0 \\ 0 & 0 & 0 & 1 \end{bmatrix},
$$

$$
\mathbf{T}_{6e} = \begin{bmatrix} -1 & 0 & 0 & -r_8 \\ 0 & 1 & 0 & 0 \\ 0 & 0 & -1 & r_7 + r_9 \\ 0 & 0 & 0 & 1 \end{bmatrix}.
\tag{1}
$$

$$
\mathbf{T}_{0e} = \mathbf{T}_{01}\mathbf{T}_{12}\mathbf{T}_{23}\mathbf{T}_{34}\mathbf{T}_{45}\mathbf{T}_{56}\mathbf{T}_{6e} = \begin{bmatrix} R_{11} & R_{12} & R_{13} & p_1 \\ R_{21} & R_{22} & R_{23} & p_2 \\ R_{31} & R_{32} & R_{33} & p_3 \\ 0 & 0 & 0 & 1 \end{bmatrix}.
\tag{2}
$$

$$
x_{RC} = p_1, \quad y_{RC} = p_2, \quad z_{RC} = p_3.
\tag{3}
$$

$$
Pan = \arctan\left(\frac{R_{32}}{R_{33}}\right), \quad Tilt = \arctan\left(\frac{-R_{31}}{\sqrt{R_{32}^2 + R_{33}^2}}\right).
\tag{4}
$$

Table 1. Link parameters of the robot manipulator.

r_1	r_2	r_3
88.5 mm	151.3 mm	403.0 mm

r_4	r_5	r_6
146.5 mm	359.0 mm	99.5 mm

r_7	r_8	r_9
83.5 mm	193.4 mm	126.0 mm

The Jacobian matrix of the differential kinematics is required to find an inverse kinematic solution. The Jacobian matrix can be derived through chain rules as shown in Equation (5), where $\mathbf{z}_i$ and $\mathbf{o}_i$ are vectors that describe the rotation axis and the origin of the ith moving frame, respectively [15]. Finally, an inverse kinematic solution for the position and orientation of the remote center can be obtained through the pseudoinverse matrix of the Jacobian matrix as shown in Equation (6).

$$
\mathbf{J} = \begin{bmatrix} \mathbf{z}_1 \times (\mathbf{o}_e - \mathbf{o}_1) & \mathbf{z}_2 \times (\mathbf{o}_e - \mathbf{o}_2) & \mathbf{z}_3 \times (\mathbf{o}_e - \mathbf{o}_3) & \mathbf{z}_4 \times (\mathbf{o}_e - \mathbf{o}_4) & \mathbf{z}_5 \times (\mathbf{o}_e - \mathbf{o}_5) & \mathbf{z}_6 \times (\mathbf{o}_e - \mathbf{o}_6) \\ \mathbf{z}_1 & \mathbf{z}_2 & \mathbf{z}_3 & \mathbf{z}_4 & \mathbf{z}_5 & \mathbf{z}_6 \end{bmatrix}
\tag{5}
$$

$$
\dot{\mathbf{q}} = \mathbf{J}^{\dagger}\dot{\mathbf{x}}
\tag{6}
$$

3. Control Strategy for Direct Teaching of Remote Center Motion

A robot manipulator is controlled based on position control to maintain a solid remote center even while the posture of the end-effector is operated with direct teaching. A force/torque sensor mounted between the tip of the cooperative robot and the end-effector is used as an input device for direct teaching, and pan and tilt motions are realized based on the remote center position according to the user's intention. A button is present on the end-effector of the surgical assistant robot, and force and moment are applied to the end-effector while the button is pressed so that the robot is driven manually. The direct teaching control of RCM is designed based on position control as shown in Figure 3. In a general position control algorithm for articulated robots, only the control input part for direct teaching is modified. The predefined remote center position and the end-effector's posture are entered into the 6-degree-of-freedom robot as a control input. The position control to solve the inverse kinematics has feedback for the actual position and orientation through the forward kinematics in the closed-loop inverse kinematics. The position control is designed by differential kinematics with PID control and back-calculation [16], and each control gain is adjusted by optimization [17]. The orientation of the end-effector is input by calculating a value generated in the force/torque sensor by direct teaching from a command generation algorithm. Through a motion generation algorithm based on impedance control, the end-effector's posture is calculated from the values generated by the force/torque sensor due to direct teaching.

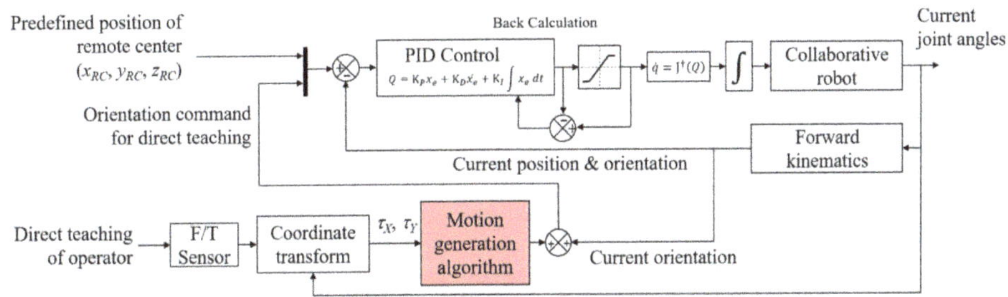

Figure 3. Block diagram of RCM control with motion generation algorithm.

3.1. Motion Generation Algorithm

When designing end-effector hand-guiding control algorithms for cooperative robots using force/torque sensors, the control model is considered a damper-mass mechanical system [18]. The torque due to the gravity force is compensated, and the robot is controlled in the velocity domain [19]. It is also important to define a proper dead zone and threshold for the values measured by the force/torque sensor [20]. Finally, the relation between the actuating variables of the input device and the robot velocity command value is designed by a motion generation algorithm [21].

The position vector from the sensor to the center of mass of the end-effector and the orientation difference between the robot base coordinate system and the sensor's coordinate are shown in Figure 4. The transformation matrix from the 6th frame to the sensor frame allows us to obtain the transformation matrix from the base to the sensor as shown in Equations (7) and (8). Because the coordinate varies depending on the robot's posture, the measured values from the 6 degrees-of-freedom force/torque sensor are converted using an adjoint matrix as shown in Equation (9) [22]. $\mathbf{f}$ and $\boldsymbol{\tau}$ are force and torque vectors, and the subscript 0 and s means the robot base frame {0} and the sensor frame {s}, respectively.

$\mathbf{R}_{0s}$ is the rotation matrix of the sensor frame in the base coordinate and $\hat{\mathbf{p}}_{0s}$ is the skew-symmetric matrix which is generated by the position vector of the sensor frame in the base coordinate.

$$\mathbf{T}_{6s} = \begin{bmatrix} -1 & 0 & 0 & 0 \\ 0 & -1 & 0 & 0 \\ 0 & 0 & 1 & r_7 \\ 0 & 0 & 0 & 1 \end{bmatrix}. \tag{7}$$

$$\mathbf{T}_{0s} = \mathbf{T}_{01}\mathbf{T}_{12}\mathbf{T}_{23}\mathbf{T}_{34}\mathbf{T}_{45}\mathbf{T}_{56}\mathbf{T}_{6s} = \begin{bmatrix} \mathbf{R}_{0s} & \mathbf{P}_{0s} \\ 0 & 1 \end{bmatrix}. \tag{8}$$

$$\begin{bmatrix} \mathbf{f}_0 \\ \boldsymbol{\tau}_0 \end{bmatrix} = \begin{bmatrix} \mathbf{R}_{0s} & 0 \\ \hat{\mathbf{P}}_{0s}\mathbf{R}_{0s} & \mathbf{R}_{0s} \end{bmatrix} \begin{bmatrix} \mathbf{f}_s \\ \boldsymbol{\tau}_s \end{bmatrix}. \tag{9}$$

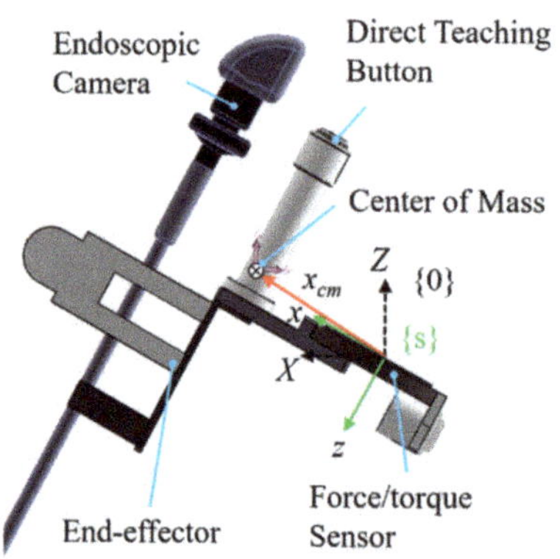

Figure 4. End-effector and sensor with moving frame and position vector of the center of mass.

Gravity-induced torque is measured on the sensor as the position of the center of gravity changes depending on the posture. It can be obtained in the robot base coordinate as shown in Equation (10). The position vector, $\mathbf{x}_{cm}$, from the sensor to the center of mass of the end-effector is $[106\text{ mm}, 0\text{ mm}, -12\text{ mm}]^T$. The mass, m, is 1.01 kg, and the gravitational acceleration, g, is 9.8 m/s^2. Finally, the input torque is derived as shown in Equation (11) as a result of compensating the torque due to gravity.

$$\begin{aligned} \mathbf{x}_g &= \mathbf{R}_{0s}\mathbf{x}_{cm}, \\ \boldsymbol{\tau}_g &= \mathbf{x}_g \times \begin{bmatrix} 0 & 0 & -mg \end{bmatrix}^T. \end{aligned} \tag{10}$$

$$\boldsymbol{\tau}_{in} = \boldsymbol{\tau}_0 - \boldsymbol{\tau}_g. \tag{11}$$

The command generation function sets the dead band $[-a, a]$ of the appropriate area so that the robot is not too sensitive to small changes in input torque and generates command values by multiplying the input torque by the proportional constant, k_1 and k_2, as shown in Figure 5. In this study, proportional constants are designed asymmetrically for sophisticated tuning of direct teaching sensitivity. The amplitude limitation of the conversion value $[-b, b]$ is also defined so that the command is not suddenly changed by a large input torque. The command value is added to the current orientation of the end-effector calculated by the forward kinematics, and the motion change by direct teaching is applied in real-time. Adding the command value for each control period of 1 ms means that the command value is the velocity of the end-effector. Based on the damper-mass system model, the input of the command generation function is input torque, and the output is motion velocity. The sensitivity of operation can be adjusted by changing the dead zone and the proportional constant. When determining the proportional constant, the target torque and the target velocity, and the limitations of the driving speed of the system should be considered in combination.

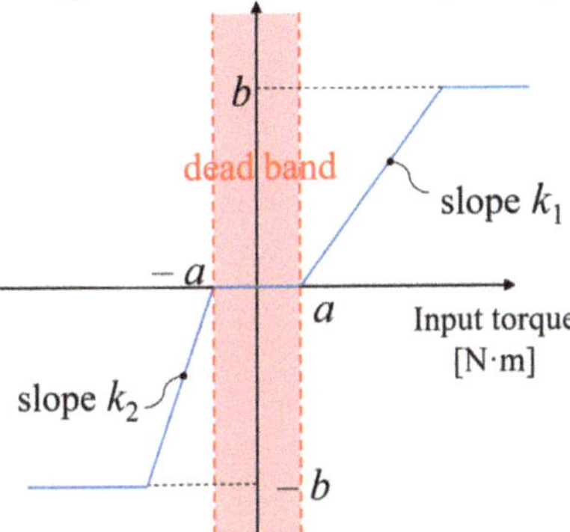

Figure 5. Relationship between input torque and robot velocity command.

3.2. Optimized Isotropic Sensitivity of Direct Teaching

The initial values of the command generation function are roughly determined among the characteristics of the system, and those are adjusted by an optimization method through a pre-test, as it is difficult to quantify the operating sensitivity of direct teaching. In general, modeling of friction terms is required when designing impedance control algorithms. However, the friction model modeled for each joint often differs from the friction model in the assembled manipulator. Although an observer is sometimes designed to estimate friction models, it increases the computation of the controller. In this work, we use a more practical approach that is simpler by tuning the command function through a pre-test. Parameter initial values of the command generation function are defined by considering the mechanical and electrical characteristics of the fabricated prototype and the desired operating sensitivity. Assuming that the dead zone range, a, is 0.2 N·m and the maximum speed, b, is 1×10^{-4} rad/ms, the slope k_1 and k_2 are determined as shown in Equation (12) so that the maximum speed can be reached with 1.5 N·m torque.

$$k_{desired} = \frac{1 \times 10^{-4}}{1.5 - 0.2} = 7.7 \times 10^{-5}. \tag{12}$$

After that, the velocity of pan/tilt motion and the generated reaction torque for arbitrary direct teaching are measured through a pre-test. The actual torque required for motion velocity is shown by contours from 18,466 data as shown in Figure 6a. The actual sensitivity of a robot differs from its intended sensitivity in amplitude and isotropy. An objective function is defined by the error norm between the actual slope and the required slope as shown in Equation (13), and the slopes of the command generation function are tuned by an optimization algorithm to minimize the objective function.

$$\tau_{opt} = \begin{cases} \tau_{mea}/\alpha, & \tau_{mea} > 0 \\ \tau_{mea}/\beta, & \tau_{mea} < 0 \end{cases},$$
$$obj\left(\alpha_{pan}, \beta_{pan}, \alpha_{tilt}, \beta_{tilt}\right) = norm\left(\frac{\sqrt{\omega_{pan}^2 + \omega_{tilt}^2}}{\sqrt{\tau_{opt,pan}^2 + \tau_{opt,tilt}^2}} - k_{desired}\right)/n, \tag{13}$$

where τ_{mea} and τ_{opt} are measured torque and optimized torque respectively, and α, β are tuning parameters to be multiplied by each slope k_i. ω is motion velocity, and n is the number of data. As a result, the error is reduced and the values of the slope are tuned as shown in Equation (14) and Table 2, resulting in the result shown in Figure 6b. Figure 6 shows the results of a verification test, and the error norm is decreased to 1.32×10^{-4} from the pre-test result.

$$k_1 = \alpha \cdot k_{initial}, \quad k_2 = \beta \cdot k_{initial}. \tag{14}$$

Table 2. Parameters of the command generation function.

Parameters		Pre-Test		Optimized Results	
	a	0.2			
	b	1×10^{-4}			
	k_1	Pan	Tilt	Pan	Tilt
		7.7×10^{-5}	7.7×10^{-5}	11.6×10^{-5}	10.5×10^{-5}
	k_2	Pan	Tilt	Pan	Tilt
		7.7×10^{-5}	7.7×10^{-5}	9.9×10^{-5}	10.5×10^{-5}
Error norm		2.01×10^{-4}		1.28×10^{-4}	

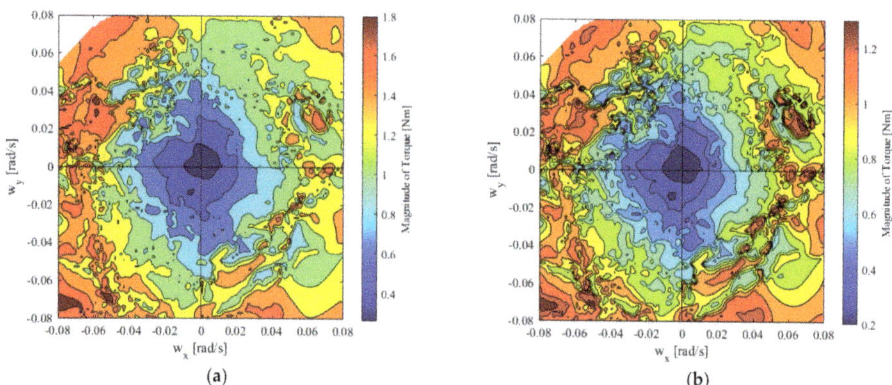

Figure 6. Contour of required torque for motion velocity: (**a**) pre-test; (**b**) optimized result.

4. Experiments

The end-effector to hold an endoscope camera is attached to the cooperative robot, and the direct teaching performance of RCM is tested as shown in Figure 7. This cooperative robot consists of driver-integrated motors (SMT-DA-series, LS Mecapion Co., Ltd., Daegu, Korea), joint modules with the Harmonic drive (MR-series, SBB Tech Co., Ltd., Gyeonggi, Korea), and a six-degree-of-freedom force/torque sensor (RFT80-6A01, ROBOTOUS INC., Gyeonggi, Korea). When the force and torque are applied to the end-effector while the teaching button is pressed, the manual traction is performed directly according to the control method shown earlier. Pan motion, Tilt motion, and mixed motion are performed manually, and the sensitivity of manipulation is investigated. The precision of the RCM is evaluated using the optical tracking system (SAKD1, DigiTrack Co., Ltd., Daegu, Korea). Orientation error is estimated through the forward kinematics based on the encoder values of the robot joints.

4.1. Pan Motion

When holding the end-effector and teaching Pan motion, torque is generated on the force/torque sensor as shown in Figure 8a, generating in the orientation command such as a dotted line of Figure 8b that is applied to the robot and followed along like a solid line. In the case of Pan motion, it can be operated with a torque of 1.10 N·m or less. During direct teaching, the remote center position error remained within 2.26 mm, with an average error of 1.52 ± 0.29 mm, as shown in Figure 8c. The orientation error according to the generated command occurs within 0.27° as shown in Figure 8d.

4.2. Tilt Motion

In the case of Tilt motion, the robot can be manipulated with torque within 0.68 N·m as shown in Figure 9a, thereby moving in the trajectory shown in Figure 9b. As shown in Figure 9c, the error of the remote center position within a maximum of 3.89 mm and an average of 1.81 ± 1.14 mm. The tilt motion follows the command within 0.18° error, as shown in Figure 9d. Due to the kinematic characteristics of articulated robots, the remote center error of Tilt motion is greater than that of Pan motion. Position errors are amplified according to joint angle errors in proportion to the distance between the remote center position and the drive joint axis. Pan motion produces major movements by joints 1, 5, and 6, and the distance between the remote center and the joint is relatively short, especially due to the dominant movement of joint 5. Tilt motion controls the major movement by joints 2, 3, and 4. The remote center position error increases compared to Pan motion because the end position error of the drive joint increases by the link length of the robot.

4.3. Pan-Tilt Mixed Motion

Holding the end effector, Pan and Tilt motion are mixed to create a cone-shaped trajectory based on the remote center. The motion of the trajectory as shown in Figure 10b is controlled within a maximum torque of 1.20 N·m as shown in Figure 10a. As shown in Figure 10c, the remote center position error of up to 4.08 mm and an average of 1.80 ± 1.03 mm is maintained for approximately 3 min of continuous operation. Orientation errors for mixed motion are within 0.70° maximum. The torque in the x-axis direction and the torque in the y-axis direction are evenly repeated with similar magnitudes for the cone-shaped rotational motion, providing a guarantee of isotropic operating sensitivity.

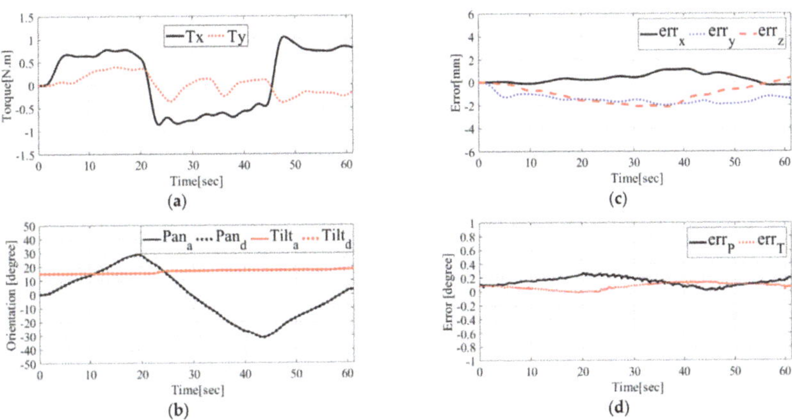

Figure 7. Experiments with a prototype of a surgical assistant robot.

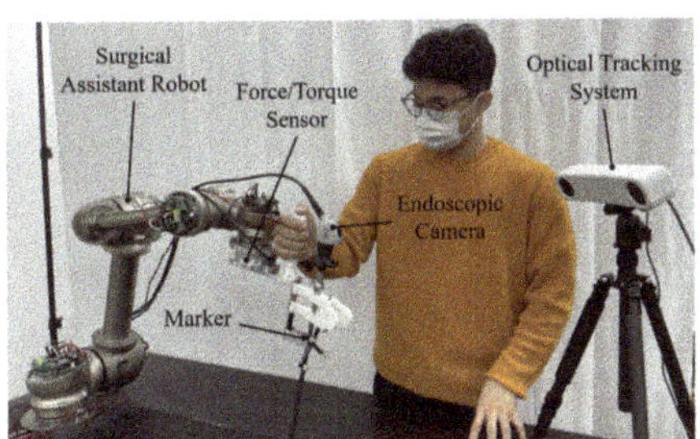

Figure 8. Experimental results in pan motion: (**a**) measured torque; (**b**) trajectory of desired and actual orientation; (**c**) position error of remote center; (**d**) orientation error of motion.

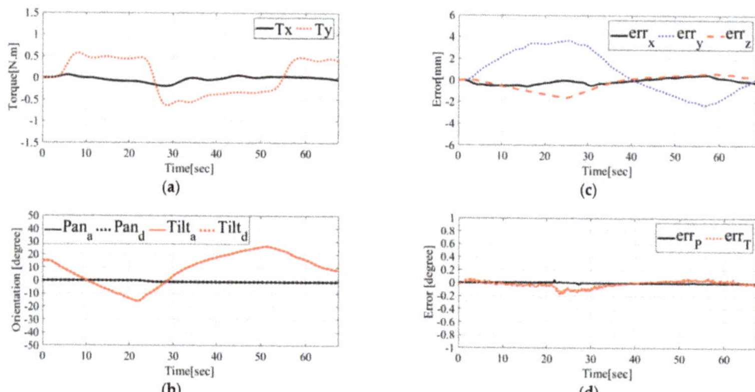

Figure 9. Experimental results in Tilt motion: (**a**) measured torque; (**b**) trajectory of desired and actual orientation; (**c**) position error of remote center; (**d**) orientation error of motion.

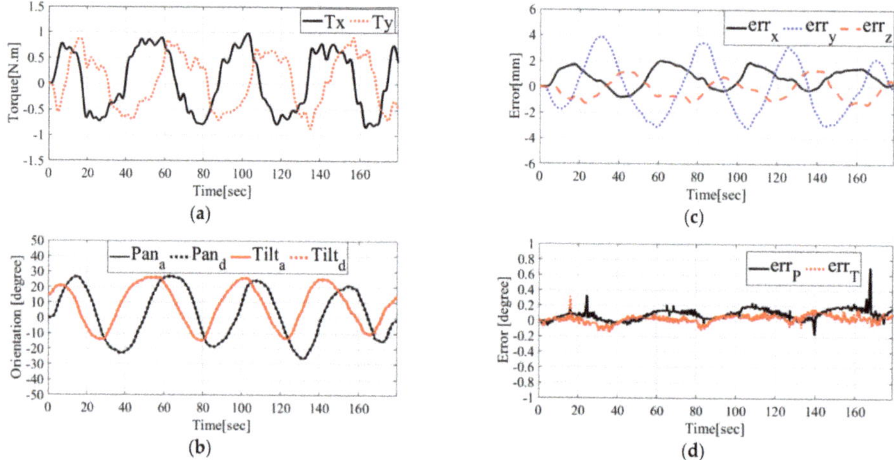

Figure 10. Experimental results in cone-shaped motion: (**a**) measured torque; (**b**) trajectory of desired and actual orientation; (**c**) position error of remote center; (**d**) orientation error of motion.

Several surgeons have surveyed the performance of this surgical assistant robot through dry testing and animal experiments and suggested that the accuracy of RCM is sufficient to be less than 5 mm for 12 mm-diameter trocars, but the operating sensitivity of direct training should be slightly improved.

5. Conclusions

This study presents a control strategy for remote center motion and direct teaching that allows a surgical assistant robot to mimic the movements of the scrub nurse's wrists. Using the proposed control method, the surgical assistant robot can be manually operated as if the surgeon holds the hand of a nurse holding the endoscope and guided it during surgery. The mechanically constrained RCM mechanism is bulky to collaborate with the surgeon on the operating table. Therefore, the conventional collaborative robot is used as the platform of the surgical assistant robot, and the precision of the RCM and the sensitivity of direct teaching are obtained simultaneously by designing the control algorithm. The position control achieves sensitive direct teaching performance by generating a posture command with the measured value from the force/torque sensor while maintaining the robust remote center position. The parameters of the command generation function are tuned through pre-test and optimization for isotropic operation sensitivity. By applying a torque within 1.20 N·m to the robot, we can hand-guide the Pan and Tilt movements of the surgical assistant robot with a remote center position error within a maximum of 4.08 mm and an average of 1.80 ± 1.03 mm. To achieve a smoother and more sensitive operation sensitivity, a more high-level command generation function will be designed and advanced techniques such as reinforcement learning will be applied to optimize parameters.

Author Contributions: Conceptualization, S.J.; methodology, S.J.; software, S.J.; validation, M.K. and Y.Z.; formal analysis, M.K.; data curation, M.K.; writing—original draft preparation, M.K.; writing—review and editing, S.J.; visualization, M.K.; supervision, S.J.; project administration, S.J.; funding acquisition, S.J. All authors have read and agreed to the published version of the manuscript.

Funding: This work was supported by the National Research Foundation of Korea (NRF) grant funded by the Korean government (MSIT) (No. NRF-2018R1C1B5043640).

Institutional Review Board Statement: Not applicable.

Informed Consent Statement: Not applicable.

Data Availability Statement: The data presented in this study are available on request from the corresponding author.

Acknowledgments: This research was conducted using equipment from Erop Co., Ltd. (Daegu, Korea).

Conflicts of Interest: The authors declare no conflict of interest.

References

1. Kim, M.; Jin, S. Study on direct teaching algorithm for remote center motion of surgical assistant robot using force/torque sensor. *J. Korea Robot. Soc.* **2020**, *15*, 309–315. [CrossRef]
2. Kuo, C.H.; Dai, J.S.; Dasgupta, P. Kinematic design considerations for minimally invasive surgical robots: An overview. *Int. J. Med. Robot. Comput. Assist. Surg.* **2012**, *8*, 127–145. [CrossRef] [PubMed]
3. Trochimczuk, R. Comparative analysis of RCM mechanisms based on parallelogram used in surgical robots for laparoscopic minimally invasive surgery. *J. Theor. Appl. Mech.* **2020**, *58*, 911–925. [CrossRef]
4. Pan, B.; Fu, Y.; Niu, G.; Xu, D. Optimization and design of remote center motion mechanism of minimally invasive surgical robotics. In Proceedings of the 11th International Conference on Ubiquitous Robots and Ambient Intelligence (URAI 2014), Kuala Lumpur, Malaysia, 12–15 November 2014. [CrossRef]
5. Shim, S.; Lee, S.; Ji, D.; Choi, H.; Hong, J. Trigonometric ratio-based remote center of motion mechanism for bone drilling. In Proceedings of the 2018 IEEE/RSJ International Conference on Intelligent Robots and Systems (IROS), Madrid, Spain, 1–5 October 2018. [CrossRef]
6. Zhu, Y.; Zhang, F. A novel remote center of motion parallel manipulator for minimally invasive celiac surgery. *Int. J. Res. Eng. Sci. (IJRES)* **2015**, *3*, 15–19.
7. Zhang, Z.; Yu, H.; Du, Z. Design and kinematic analysis of a parallel robot with remote center of motion for minimally invasive surgery. In Proceedings of the 2015 IEEE International Conference on Mechatronics and Automation (ICMA), Beijing, China, 2–5 August 2015. [CrossRef]
8. Dombre, E.; Michelin, M.; Pierrot, F.; Poignet, P.; Bidaud, P.; Morel, G.; Ortmaier, T.; Sallé, D.; Zemiti, N.; Gravez, P.; et al. MARGE Project: Design, modeling and control of assistive devices for minimally invasive surgery. In Proceedings of the Medical Image Computing and Computer-Assisted Intervention (MICCAI 2004), Saint-Malo, France, 26–29 September 2004. [CrossRef]
9. Lee, S.D.; Ahn, K.H.; Song, J.B. Torque control based sensorless hand guiding for direct robot teaching. In Proceedings of the 2016 IEEE/RSJ International Conference on Intelligent Robots and Systems (IROS), Daejeon, Korea, 9–14 October 2016. [CrossRef]
10. Ahn, K.H.; Song, J.B. Cartesian space direct teaching for intuitive teaching of a sensorless collaborative robot. *J. Korea Robot. Soc.* **2019**, *14*, 311–317. [CrossRef]
11. Albu-Schäffer, A.; Haddadin, S.; Ott, C.; Stemmer, A.; Wimböck, T.; Hirzinger, G. The DLR lightweight robot: Design and control concepts for robots in human environments. *Ind. Robot Int. J. Robot. Res. Appl.* **2007**, *34*, 376–385. [CrossRef]
12. Park, C.; Kyung, J.H.; Do, H.M.; Choi, T. Development of direct teaching robot system. In Proceedings of the 8th International Conference on Ubiquitous Robots and Ambient Intelligence (URAI 2011), Incheon, Korea, 23–26 November 2011. [CrossRef]
13. Ahn, J.H.; Kang, S.; Cho, C.; Hwang, J.; Suh, M. Design of robot direct-teaching tool and its application to path generation for die induction hardening. In Proceedings of the 2002 International Conference on Control, Automation and Systems (ICCAS), Jeonbuk, Korea, 16–19 October 2002.

14. Kim, H.J.; Back, J.H.; Song, J.B. Direct teaching and playback algorithm for peg-in-hole task using impedance control. *J. Inst. Control Robot. Syst.* **2009**, *15*, 538–542. [CrossRef]
15. Spong, M.W.; Hutchinson, S.; Vidyasagar, M. *Robot Modeling and Control*, 1st ed.; John Wiley & Sons, Inc.: New York, NY, USA, 2006; pp. 113–130.
16. Jin, S.; Lee, S.K.; Lee, J.; Han, S. Kinematic model and real-time path generator for a wire-driven surgical robot arm with articulated joint structure. *Appl. Sci.* **2019**, *9*, 4114. [CrossRef]
17. Jin, S.; Han, S. Gain optimization of kinematic control for wire-driven surgical robot with layered joint structure considering actuation velocity bound. *J. Korea Robot. Soc.* **2020**, *15*, 212–220. [CrossRef]
18. Safeea, M.; Bearee, R.; Neto, P. End-Effector precise hand-guiding for collaborative robots. In *ROBOT 2017: Third Iberian Robotics Conference*; Springer: Cham, Switzerland, 2017; Volume 694, pp. 595–605. [CrossRef]
19. Massa, D.; Callegari, M.; Cristalli, C. Manual guidance for industrial robot programming. *Ind. Robot* **2015**, *42*, 457–465. [CrossRef]
20. Rodamilans, G.B.; Villani, E.; Trabasso, L.G.; Oliveira, W.R.; Suterio, R. A comparison of industrial robots interface: Force guidance system and teach pendant operation. *Ind. Robot* **2016**, *43*, 552–562. [CrossRef]
21. Fujii, M.; Murakami, H.; Sonehara, M. Study on application of a human-robot collaborative system using hand-guiding in a production line. *IHI Eng. Rev.* **2016**, *49*, 24–29.
22. Murray, R.M.; Li, Z.; Sastry, S.S. *A Mathematical Introduction to Robotic Manipulation*, 1st ed.; CRC Press: Boca Raton, FL, USA, 1994; pp. 61–69.

applied sciences

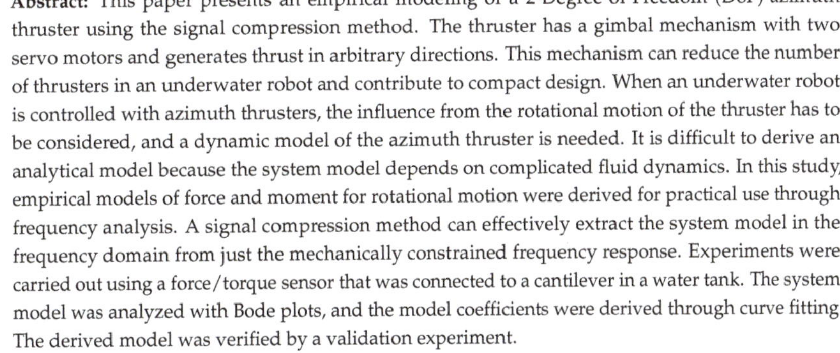

Article

Empirical Modeling of 2-Degree-of-Freedom Azimuth Underwater Thruster Using a Signal Compression Method

Cheol-Su Jeong [1], Gunwoo Kim [2], Inwon Lee [3] and Sangrok Jin [4,*]

1 Research Institute of Mechanical Technology (RIMT), Pusan National University, Pusan 46241, Korea; tk0083@nate.com
2 Division of Robotics Convergence, Pusan National University, Pusan 46241, Korea; gwkim@pusan.ac.kr
3 Department of Naval Architecture and Ocean Engineering, Pusan National University, Pusan 46241, Korea; inwon@pusan.ac.kr
4 School of Mechanical Engineering, Pusan National University, Pusan 46241, Korea
* Correspondence: rokjin17@pusan.ac.kr; Tel.: +82-51-510-2984

Featured Application: Underwater robot.

Abstract: This paper presents an empirical modeling of a 2-Degree-of-Freedom (DoF) azimuth thruster using the signal compression method. The thruster has a gimbal mechanism with two servo motors and generates thrust in arbitrary directions. This mechanism can reduce the number of thrusters in an underwater robot and contribute to compact design. When an underwater robot is controlled with azimuth thrusters, the influence from the rotational motion of the thruster has to be considered, and a dynamic model of the azimuth thruster is needed. It is difficult to derive an analytical model because the system model depends on complicated fluid dynamics. In this study, empirical models of force and moment for rotational motion were derived for practical use through frequency analysis. A signal compression method can effectively extract the system model in the frequency domain from just the mechanically constrained frequency response. Experiments were carried out using a force/torque sensor that was connected to a cantilever in a water tank. The system model was analyzed with Bode plots, and the model coefficients were derived through curve fitting. The derived model was verified by a validation experiment.

Keywords: azimuth thruster; thruster modeling; signal compression method; frequency response analysis; empirical modeling

1. Introduction

A typical mechanism for the propulsion of underwater robots is a thruster with a propeller. The dynamic model of a tilting thruster is essential for designing a controller for underwater robots. Due to the complex effects of nonlinear fluid dynamics, studies have derived the dynamic model using experiments. Yerger suggested that the thrust force is proportional to the signed square of the propeller's rotational speed [1]. Healy proposed a nonlinear model considering the fluid dynamics of the motor model and propeller blade [2]. Bachmayer presented a combination model consisting of an axial flow model and a rotational flow model. [3]. Blanke developed a three-state model based on dimensionless propeller parameters, thrust coefficients, and an advance ratio [4]. Kim proposed an advanced three-state model considering the surrounding flow rate and inflow angle [5–7]. Yu [8] and Spearenberg [9] numerically analyzed the effect of thrust ducts on different boundary conditions. Song conducted Bollard-pull experiments to find the relationship between axial thrust and input while considering the effect of thrust–thrust and thrust–hull interactions [10]. Boem used a system identification process to derive an accurate dynamic model for commercially available fixed-pitch variable speed thrusters [11]. Odetti devel-

117

oped a novel thruster based on a pump-jet [12] and applied it to ASV (Autonomous Surface Vehicle) in shallow water [13].

Most underwater robots are driven by fixed thrusters, but tilting thrusters have recently been used in underwater robots [14]. A tilting thruster can reduce the number of thrusters in an underwater robot and contributes to compact design [15]. When an underwater robot is controlled with tilting thrusters, the influence of the rotational motion of the thruster has to be considered, unlike a fixed thruster. The rotating thrust force can be calculated by multiplying the steady-state thrust force by a trigonometric function of the rotation angle [16].

When a thruster is operated with rotation motion, additional forces and moments are generated, and they can have a complicated relationship. Studies have numerically analyzed dynamic models according to the rotation angle of the azimuth thrusters [17,18], but a model that takes into account the effects that occur while the thruster is rotating is needed. In our previous research, dynamic models for rotation motion were derived by oscillating a 1-Degree-of-Freedom (DoF) tilting thruster using sine waves [19]. Modeling errors occurred as a result of relying on discrete data in the frequency domain, and it was necessary to design a model-free robust controller to overcome them [20].

This paper presents a 2-DoF azimuth thruster, as shown in Figure 1. Like fish fins, the thruster has a 2-DoF direction of propulsion. There are two rotational joints with the gimbal mechanism. It can produce effects similar to those of bioinspired robots that produce two-way propulsion through tail motion [21]. In a steady state, the direction of the thrust is the same as the direction of the thruster. However, the thrust is bent and dispersed during the rotational motion of the azimuth thruster, and a reaction moment also occurs. It is difficult to analytically describe this phenomenon, because the system model highly depends on complicated fluid dynamics. In this study, a signal compression method was applied to derive an empirical model of the azimuth thruster. Signal compression methods are often used for the empirical modeling of mechanical systems and can extract the frequency response of a system in the whole range of mechanically constrained frequencies using a pseudo-impulse signal [22]. Signal compression methods have been used for the cylinder modeling of an automatic excavator [23] and friction modeling of a robot manipulator [24].

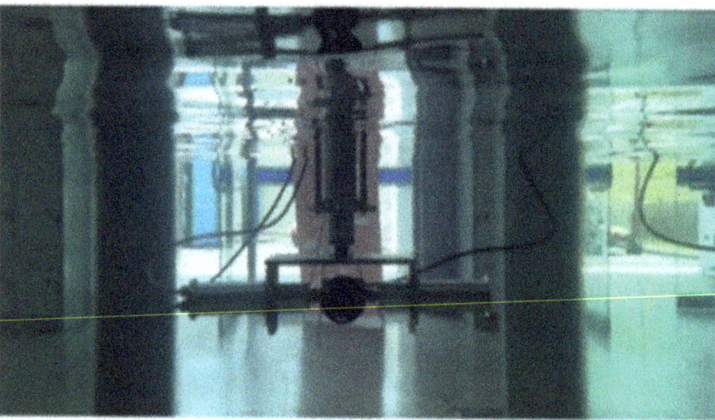

Figure 1. Prototype of 2-Degree-of-Freedom (DoF) azimuth thruster.

In repeated experiments, the velocity profile of the pseudo-impulse input was applied to the two rotational axes, and the frequency response of the longitudinal thrust force, lateral thrust force, and reaction moment were analyzed with a Bode plot. Model expressions were derived through a curve-fitting technique. The validity of the derived model was verified by comparing the experimental results with model-based simulation values. This work provides references on how to derive empirical models of vectoring thrusters that can be applied to a variety of underwater robots.

The remainder of the paper is organized as follows. The prototype of the 2-DoF azimuth thruster and the experimental apparatus are described in Section 2. The signal compression method for the empirical modeling is introduced in Section 3. In Section 4, the empirical models of thrust force and reaction moment are presented with an analysis of the Bode plot of the pseudo-impulse response. Section 5 shows a comparison between the model and experimental results. The validity of the empirical model is also discussed. Our conclusions are given in Section 6.

2. Description of 2-DoF Azimuth Thruster

A prototype of a 2-DoF azimuth thruster was fabricated, as shown in Figure 2. The system consists of a thruster and two servo motors and has a gimbal mechanism for 2-DoF rotational motion. A counter mass is used so that the center of mass and the rotational axis intersect. The longitudinal thrust that is generated by the thruster and the tilting angles of the two servo motors are system inputs, and the actual thrust force and reaction torque are outputs of the system model. The thruster (RCD-MI60, 24VDC, RHINCODON, China) has a maximum thrust force of 5.5 kgf. The duct diameter is 95 mm, and the weight in air is 850 g. The gimbal mechanism is actuated by two BLDC (Brushless DC) motors (EC-max 30, 24VDC, Maxon, Switzerland). The total weight of azimuth thruster system is 8.56 kg.

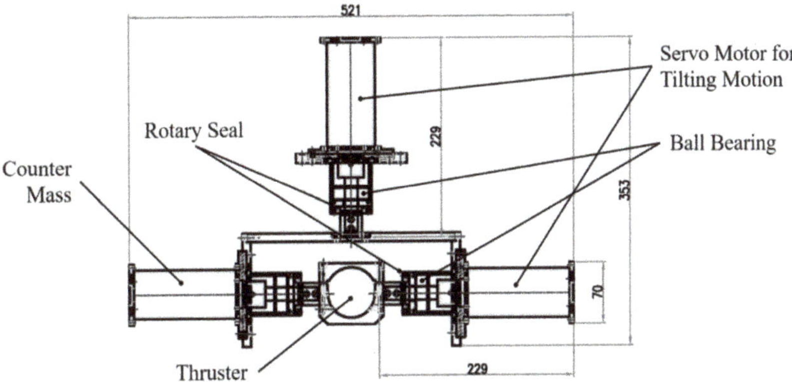

Figure 2. Design drawing of 2-DoF azimuth thruster.

The generated thrust force and reaction moment are shown in Figure 3. The X-Y-Z frame is a fixed frame, and the x-y-z frame is a moving frame that is attached to the tilting thruster. The thrust force is bent and dispersed due to rotational motion, and it can be decomposed into f_x, f_y, and f_z in the moving frame. These forces can be converted to f_X, f_Y, and f_Z in the fixed frame using a rotation matrix. In this study, the frequency response of forces was analyzed in the moving frame. The reaction moment was analyzed within the framework of the gimbal mechanism. The reaction moment M_Z generated by the rotational motion of the 1st axis is defined in the fixed frame, and the reaction moment M_y for the rotational motion of the 2nd axis is defined in the moving frame.

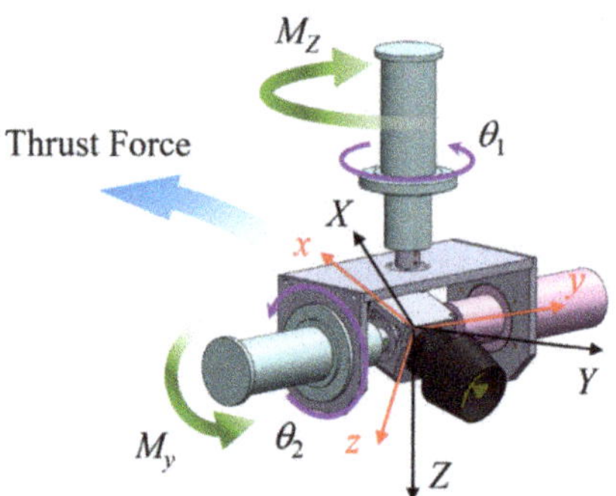

Figure 3. Free body diagram of thruster force and reaction moment.

A diagram of the experimental apparatus is shown in Figure 4. In order to measure forces and moments, the 2-DoF azimuth thruster unit was installed at the end of a cantilever with a 6-DoF force/torque sensor (RFT80-6A01, ROBOTUS, Gyeonggi, Korea) attached at the clamped end. The thruster is driven by a pulse-width modulation (PWM) signal with a range of 55–95%, and the BLDC motors are driven by position controllers (EPOS2 25/2, Maxon, Switzerland). A real-time controller (NI CompactRIO, National Instruments Corp., Austin, TX, USA) controls the actuation and receives the values from the force/torque sensor. The measured forces and torques from the sensor are converted to the values of the system origin through the adjoint matrix of the frame transformation.

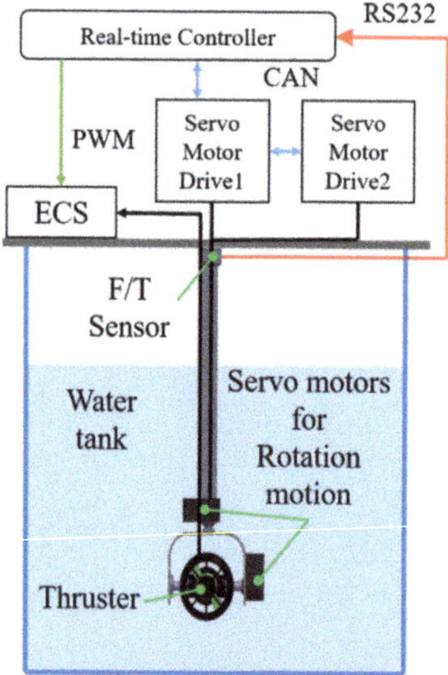

Figure 4. Diagram of the experimental apparatus.

3. Signal Compression Method

Signal compression methods are useful tools for the frequency analysis of mechanical systems under physical limitations. The impulse response can show the dynamic characteristics of a system in the whole range of frequency. However, many mechanical systems cannot make an ideal impulse input. In the signal compression method, a pseudo-impulse input that is expanded with phase delay is applied to the system. The output that is obtained from the pseudo-impulse input is mathematically equivalent to the impulse response. After the equivalent output is compressed, the frequency response of the dynamic system can be observed within the desired frequency range. A diagram of the signal compression method is shown in Figure 5. In this study, the frequency response was analyzed for system modeling with the Bode plot before transformation into the time domain.

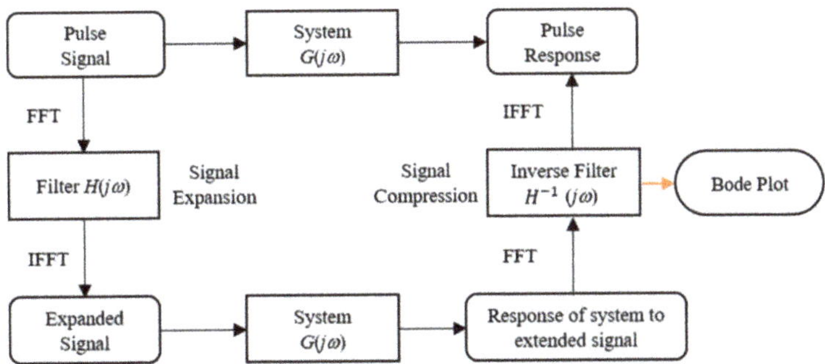

Figure 5. Diagram of signal compression method.

The dynamic characteristics due to rotational motion are affected by the angular velocity, acceleration, and higher-order terms. In this study, the angular velocity was input to observe the system characteristics. The power spectrum of ideal impulse input is constant in the whole range of frequency. The power spectrum within the desired frequency was designed as shown in Figure 6a, considering the capability of servo motors. It can be derived by a function of the power spectral density as follows:

$$P(n) = Q \exp\left\{-\left(\frac{n}{a}\right)^{12}\right\}, \tag{1}$$

where Q and a are parameters, and n is the number of data values.

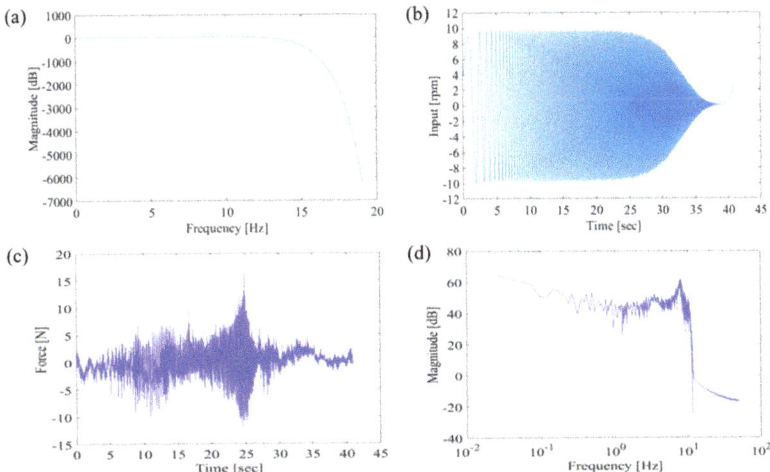

Figure 6. Input and output in the signal compression method. (**a**) Power spectrum within desired frequency; (**b**) pseudo-impulse signal input; (**c**) output for pseudo-impulse signal; (**d**) Bode plot of signal compression result.

The range of frequency of the power spectrum is manipulated using the parameter a. In this study, a was set as 450, and the range of frequency was designed as up to 10 Hz, as shown in Figure 6a, considering the capability of the servo motors. The pseudo-impulse input was derived by signal expansion as follows:

$$X(n) + jY(n) = \begin{cases} P(n)H(jn), & 0 \leq n \leq \frac{N}{2} - 1 \\ 0, & n = \frac{N}{2} \\ X(N-n) + jY(N-n), & \frac{N}{2} + 1 \leq n \leq N - 1 \end{cases} \tag{2}$$

$$H(jn) = \exp\left\{ -\left(\frac{12n^2}{b}\right)j \right\}. \tag{3}$$

$H(jn)$ is a function of the filter for signal expansion with a time delay. The time delay depends on the parameter b, which is 2100 in this case. Through the inverse fast Fourier transform (IFFT), the expanded pseudo-impulse signal can be obtained in the time domain as shown in Figure 6b, and it is applied to the system. The equivalent output is generated as shown in Figure 6c, and then the impulse response can be obtained via signal compression with an inverse filter. Finally, the Bode plot shows the frequency response of the system, as shown in Figure 6d.

4. Empirical Modeling

Variations of the thrust force and moment due to the rotational motion of the azimuth thruster were modeled based on the steady-state thrust force. Longitudinal thrust force is generated by a thruster, as shown in Figure 7a. The steady-state thrust force is proportional to the duty ratio of the PWM from the motor drive, as shown in Figure 7b. The empirical model expression of the steady-state thrust force for PWM is as follows:

$$f_{ss} = 0.55\text{PWM} - 7.10. \tag{4}$$

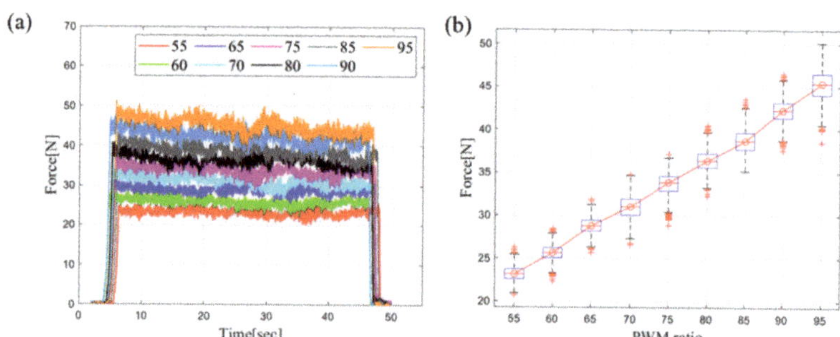

Figure 7. The longitudinal thrust force according to pulse-width modulation (PWM). (**a**) Thrust force of a fixed thruster in time domain for PWM; (**b**) steady-state thrust force with PWM variation.

The steps of empirical modeling are as follows: (1) Repeat the rotational motion experiment, increasing the PWM by 5% from 55% to 95%. (2) Measure the force and moment by entering the velocity profile as shown in Figure 6b on the 1st and 2nd axes, respectively. (3) Calculate the force of the moving frame reference according to the angle of each axis using the adjoint matrix. (4) Draw the Bode plot of the pseudo-impulse response through the signal compression. (5) Derive the model through the curve fitting with the numerical method to maximize the correlation number.

As a result of the repeated experiments, waveforms for x-direction attenuation and y- and z-direction generation of propulsion were similar, and the reaction moments showed different forms of waveforms. The models can be derived by the analysis of the Bode plot with the corner frequency and the slope of asymptotes [25], as shown in Figure 8. As the phase plot is noisy, we focused on the magnitude plot. The model expressions of the force and moment are defined as follows:

$$\frac{F(s)}{\dot{\theta}(s)} = \frac{K(s^2 + as + b)}{(s^2 + cs + d)(s^2 + es + g)} \tag{5}$$

$$\frac{M(s)}{\dot{\theta}(s)} = \frac{Ks^2}{s^4 + as^3 + bs^2 + cs + d} \tag{6}$$

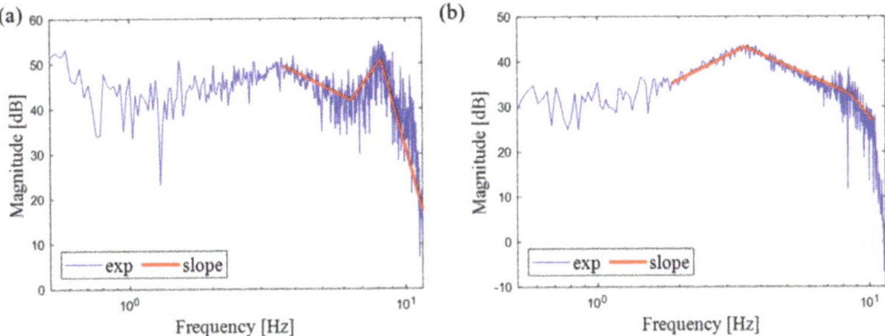

Figure 8. Magnitude Bode plot and slope of asymptote: (**a**) force; (**b**) moment.

The force and moment models have the same system order but slightly different equations. The coefficients of the model expression were obtained through numerical optimization that maximizes the correlation coefficients. The results of experiments with significantly different tendencies through repeated experiments were eliminated. The objective function of modeling is defined as the maximum correlation coefficient with repeated experimental results, and uncertainty in the data is handled.

4.1. Model for 1st-Axis Rotation Motion

When we placed the velocity profile of the pseudo-impulse input on the 1st rotational axis, the compressed response showed variation of the longitudinal thrust force in the x-direction, the lateral thrust force in the y-direction, and the reaction moment in the Z-direction. For each PWM, curve fitting was performed for three or more experimental values to extract the results with the highest correlation number, as shown in Tables 1–3. As an example, Figure 9 shows the Bode plot of the experimental values and the proposed model of $F_{1x}(s)$ and $F_{1y}(s)$ at 75% PWM.

Table 1. Model coefficient of $F_{1x}(s)$ in 1st-axis rotation.

PWM	K	a	b	c	d	e	g	Correlation Number
55	1018.79	97.21	138.87	18.70	0.48	7.30	3410.18	0.54
60	982.72	88.35	143.12	12.27	1.00	4.15	3340.46	0.65
65	1146.46	102.21	139.95	16.34	1.00	5.27	3418.75	0.58
70	1006.43	80.21	124.82	11.13	1.00	3.36	3073.84	0.69
75	1035.34	76.51	127.9	9.88	1.00	6.40	3228.49	0.56
80	1105.16	106.39	132.55	14.49	1.00	4.76	3375.93	0.63
85	1365.23	96.06	141.85	9.13	1.70	5.82	3399.70	0.58
90	1060.66	101.21	129.96	8.24	1.00	6.69	3309.18	0.55
95	985.19	95.17	132.58	8.24	1.00	5.32	3274.37	0.60

Table 2. Model coefficient of $F_{1y}(s)$ in 1st-axis rotation.

PWM	K	a	b	c	d	e	g	Correlation Number
55	846.72	195.17	115.87	31.70	25.03	4.01	2453.09	0.73
60	771.28	104.22	105.56	11.93	52.96	1.79	2650.87	0.80
65	825.43	210.56	112.90	26.99	63.49	3.43	2555.52	0.74
70	901.74	182.38	110.55	22.53	131.94	2.72	2569.85	0.79
75	837.14	169.98	112.50	20.53	125.91	3.93	2440.16	0.76
80	863.48	205.88	114.20	21.99	58.22	3.82	2573.51	0.76
85	1059.66	173.98	115.78	16.76	28.44	4.68	2559.41	0.71
90	941.98	184.16	116.35	15.19	13.64	3.56	2524.77	0.75
95	1079.87	177.30	116.02	18.97	24.50	5.08	2450.76	0.71

Table 3. Model coefficient of $M_{1z}(s)$ in 1st-axis rotation.

PWM	K	a	b	c	d	Correlation Number
55	685.06	14.85	864.03	4751.23	229.86	0.81
60	677.55	15.05	851.00	4751.25	229.90	0.82
65	694.68	14.89	899.98	4750.00	230.00	0.84
70	700.19	14.51	773.18	4750.79	230.22	0.82
75	711.34	15.20	839.31	4751.00	229.90	0.84
80	694.65	14.90	899.99	4750.00	230.00	0.81
85	659.16	15.05	838.37	4750.60	230.07	0.84
90	606.12	13.16	827.03	4750.34	230.12	0.80
95	681.94	15.01	848.25	4751.01	229.96	0.84

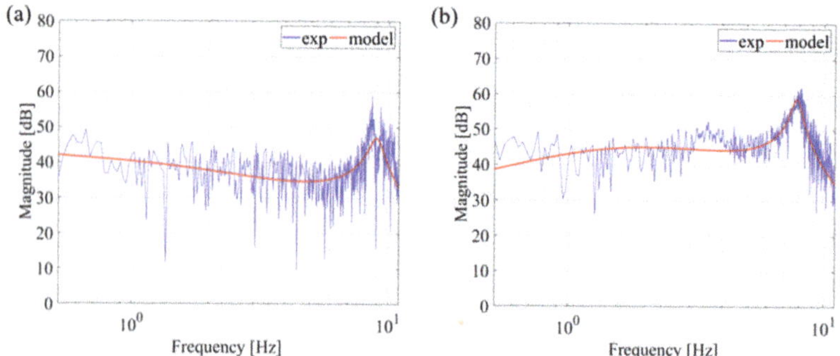

Figure 9. Magnitude Bode plot of experimental results and model in 1st-axis rotation with 75% PWM: (**a**) $F_{1x}(s)$; (**b**) $F_{1y}(s)$.

The variations in each parameter for PWM allowed us to analyze the characteristics of the model, as shown in Figure 10. The average correlation number of $F_{1x}(s)$ is 0.60. It is not a high value but is within a reasonable range. The gain K increases with increasing propulsion. Differences in parameters d and g indicate significant differences in the natural frequency of the longitudinal thrust force and the lateral force.

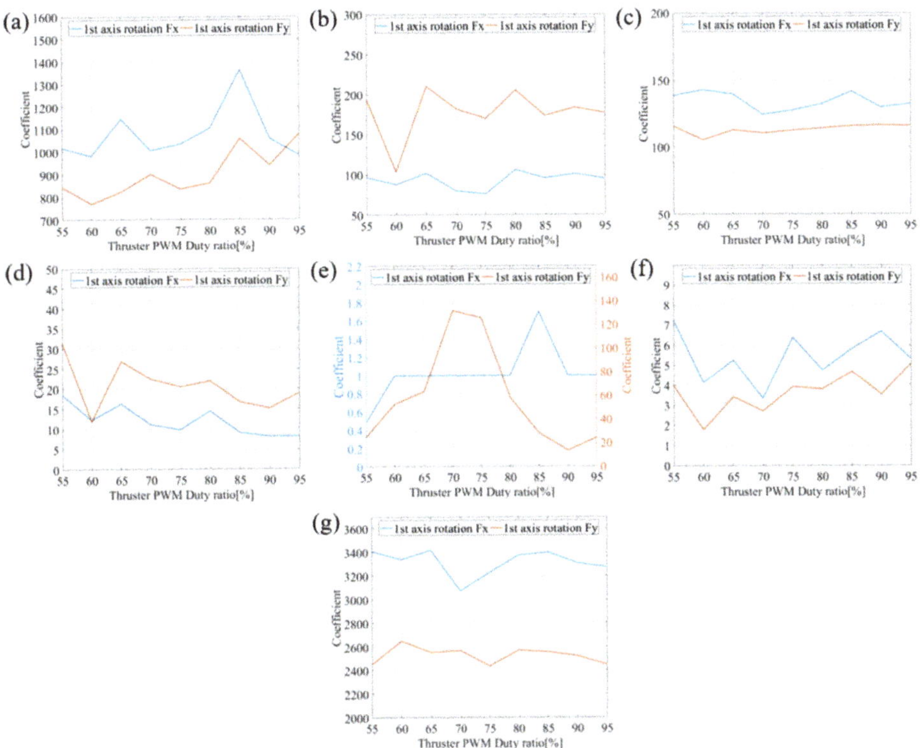

Figure 10. Model coefficient of thrust force in 1st-axis rotation according to PWM: (**a**) *K*, (**b**) *a*, (**c**) *b*, (**d**) *c*, (**e**) *d*, (**f**) *e*, and (**g**) *g*.

Figure 11 shows the Bode plot of the experimental values and the proposed model of $M_{1Z}(s)$ at 75% PWM. The correlation number of the moment model is 0.80 or higher. The empirical model matches the experimental results very well. The variations in each parameter for PWM are shown in Figure 12. The parameters of the model are not significantly affected by the change in thrust force. The structure of the gimbal is a more dominant factor in the reaction moment model than in the change in thrust force.

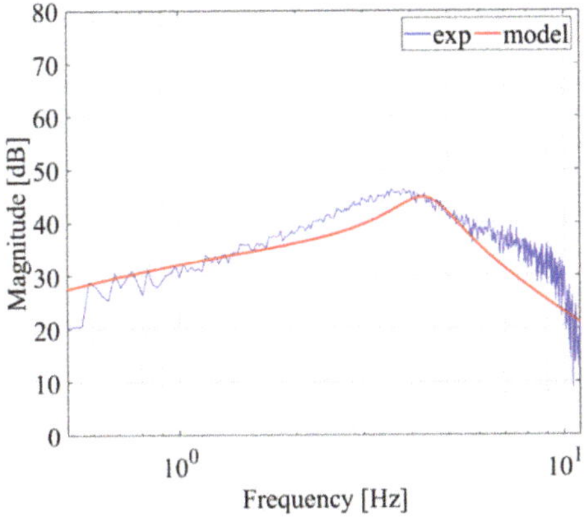

Figure 11. Magnitude Bode plot of experimental results and model of $M_{1z}(s)$ in 1st-axis rotation with 75% PWM.

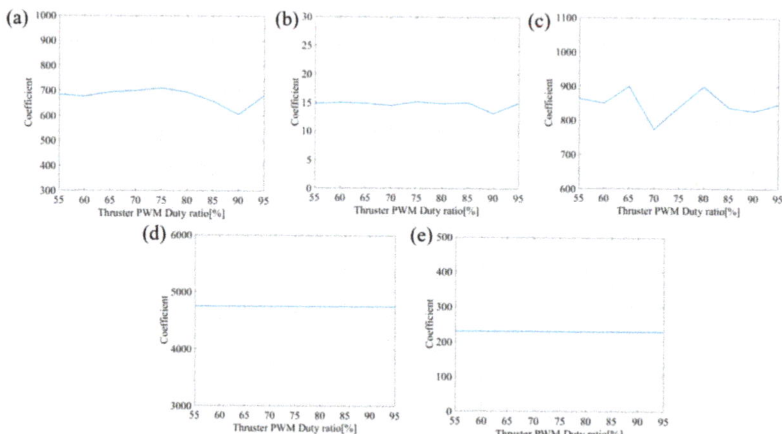

Figure 12. Model coefficient of moment in 1st-axis rotation according to PWM: (**a**) K, (**b**) a, (**c**) b, (**d**) c, and (**e**) d.

4.2. Model for 2nd-Axis Rotation Motion

When we placed the velocity profile of the pseudo-impulse input on the 2nd rotational axis, the compressed response showed the variation of the longitudinal thrust force in the x-direction, the lateral thrust force in the z-direction, and the reaction moment in the y-direction. The 2nd-axis rotational motion did not rotate the gimbal structure, unlike the 1st-axis rotational motion. When the moment was measured at the end of the cantilever, the moment in the y-direction was mixed with the force in the x-direction. It was too small to separate a meaningful value from the measurement. In this study, the reaction moment in the y-direction was neglected.

The model parameters for each PWM are shown in Tables 4 and 5. As an example, Figure 13 shows the Bode plot of the experimental values and the proposed model of $F_{2x}(s)$ and $F_{2z}(s)$ at 75% PWM. The tendency of the model parameters according to PWM was analyzed, as shown in Figure 14. Like in the 1st-axis motion, the gain K increases with increasing thrust. The model for the change in thrust forces differs somewhat from the 1st-axis motion. It can be seen that the movement of the gimbal structure has a considerable influence on the change in propulsion.

Table 4. Model coefficient of $F_{2x}(s)$ in 2nd-axis rotation.

PWM	K	a	b	c	d	e	g	Correlation Number
55	846.72	195.17	115.87	31.70	25.03	4.01	2453.09	0.73
60	771.28	104.22	105.56	11.93	52.96	1.79	2650.87	0.80
65	825.43	210.56	112.90	26.99	63.49	3.43	2555.52	0.74
70	901.74	182.38	110.55	22.53	131.94	2.72	2569.85	0.79
75	837.14	169.98	112.50	20.53	125.91	3.93	2440.16	0.76
80	863.48	205.88	114.20	21.99	58.22	3.82	2573.51	0.76
85	1059.66	173.98	115.78	16.76	28.44	4.68	2559.41	0.71
90	941.98	184.16	116.35	15.19	13.64	3.56	2524.77	0.75
95	1079.87	177.30	116.02	18.97	24.50	5.08	2450.76	0.71

Table 5. Model coefficient of $F_{2z}(s)$ in 2nd-axis rotation.

PWM	K	a	b	c	d	e	g	Correlation Number
55	1289.06	51.75	120.17	6.19	11.49	8.10	2712.26	0.66
60	1382.06	49.87	119.74	5.31	19.14	6.83	2713.70	0.69
65	1456.32	59.57	120.01	5.57	11.93	7.78	2712.20	0.64
70	1483.13	60.73	119.41	5.05	20.00	6.61	2663.02	0.69
75	1357.27	74.79	119.92	4.59	9.89	6.85	2712.36	0.60
80	1705.72	75.73	119.95	4.49	17.88	7.94	2815.79	0.67
85	1653.80	70.92	119.85	4.65	13.14	5.84	2711.62	0.68
90	1829.85	70.42	119.87	2.68	11.27	5.32	2715.95	0.72
95	2029.13	95.62	117.50	3.30	19.81	7.25	2820.89	0.71

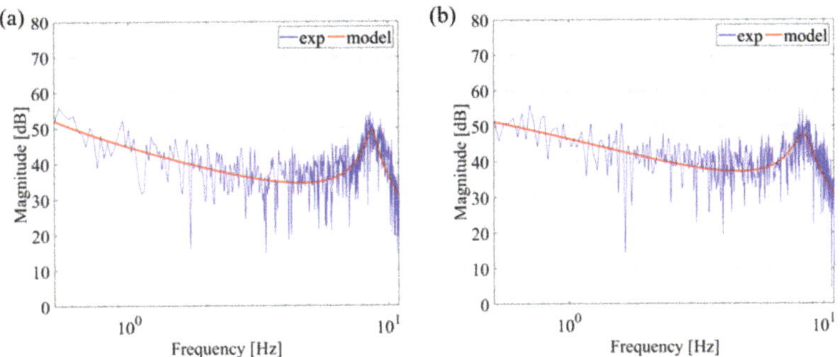

Figure 13. Magnitude Bode plot of experimental results and model in 2nd-axis rotation with 75% PWM: (**a**) $F_{2x}(s)$; (**b**) $F_{2z}(s)$.

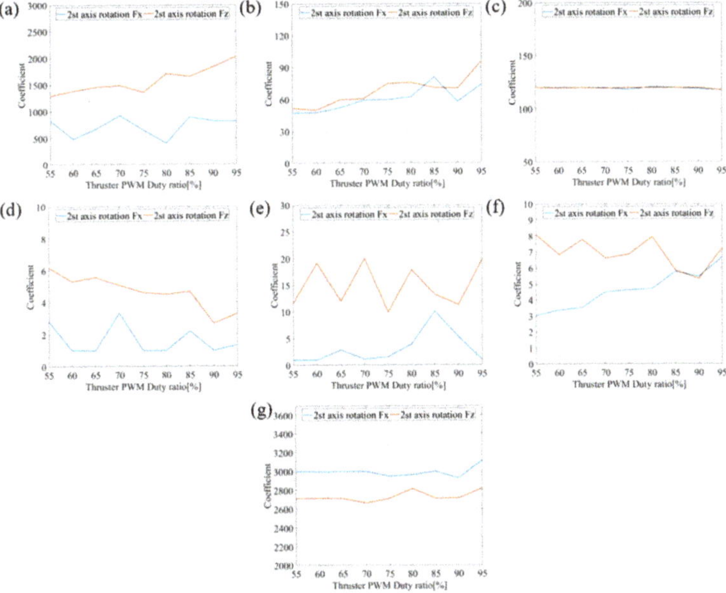

Figure 14. Model coefficient of thrust force in 2nd-axis rotation according to PWM: (**a**) K, (**b**) a, (**c**) b, (**d**) c, (**e**) d, (**f**) e, and (**g**) g.

5. Comparison between Model and Experimental Results

To validate the feasibility of the derived model, we compared the experimental result for an arbitrary rotation path with the model-based simulation value. It is recommended that the longitudinal and lateral thrust force be interpreted on a moving frame basis for dynamic interpretation, but as an application, the forces of the azimuth thruster are converted into forces in the fixed frame based on the location of the attachment. We derived the forces and moments in the moving frame by entering the inputs applied in experiments into a simulation model on two axes. The forces were converted into the fixed coordinates through a rotational matrix. The Z-axis moment was directly compared with the experimental results without coordinate transformation. The simulation was performed in MATLAB (Mathworks Inc., Natick, MA, USA).

5.1. Verification Test

A verification experiment was conducted by entering commands such as step inputs of different angles on the two rotational axes. The actual rotation angle for the step input is shown in Figure 15. The first axis rotated 45°, and the second axis rotated 30°. The thruster was operated in advance with 65% PWM to start the rotational motion in a steady state. First, the initial offset due to gravity and disturbance was calibrated, and then the forces and torque were measured.

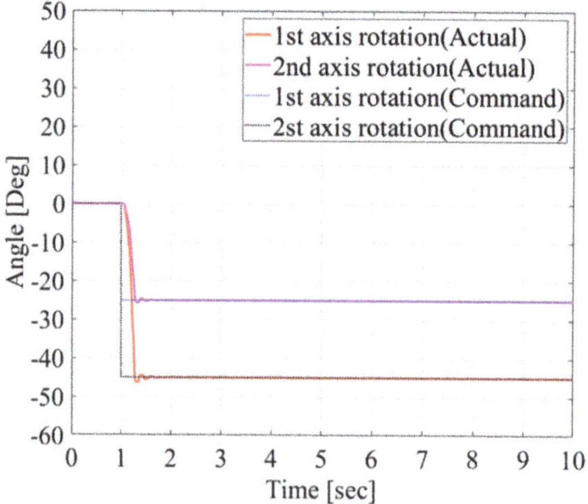

Figure 15. Angular position input for the verification test.

5.2. Simulation Based on the Proposed Model

As in the experiment, we simulated the response to a step input. Since the derived models are for rotational speed, they were transformed into models for the angle of rotation as follows:

$$\frac{F(s)}{\theta(s)} = \frac{Ks\left(s^2 + as + b\right)}{\left(s^2 + cs + d\right)\left(s^2 + es + g\right)} \tag{7}$$

$$\frac{M(s)}{\theta(s)} = \frac{Ks^3}{s^4 + as^3 + bs^2 + cs + d} \tag{8}$$

Assuming that the forces due to the rotation of each axis are independent of each other, the longitudinal thrust was calculated by superposing the change in axial force of each axis onto the steady-state force as follows:

$$f_x(t) = f_{ss} + \left\{f_{1x}(t) + f_{2x}(t)\right\} \tag{9}$$

The sum of the models of the derived *x*-axis force variation allowed us to obtain the attenuated *x*-axis thrust force. The rotation of the coordinate system by two axes is expressed in a matrix as follows:

$$R(\theta_1, \theta_2) = R_z(\theta_1)R_y(\theta_2) = \begin{bmatrix} \cos\theta_1\cos\theta_2 & -\sin\theta_1 & \cos\theta_1\sin\theta_2 \\ \sin\theta_1\cos\theta_2 & \cos\theta_1 & \sin\theta_1\sin\theta_2 \\ -\sin\theta_2 & 0 & \cos\theta_2 \end{bmatrix} \tag{10}$$

The results of force simulation in moving coordinates can be transformed into forces in fixed coordinates as follows:

$$\begin{bmatrix} f_X \\ f_Y \\ f_Z \end{bmatrix} = R^T(\theta_1, \theta_2) \begin{bmatrix} f_x \\ f_{1y} \\ f_{2z} \end{bmatrix} \tag{11}$$

The experimental results and simulated values are compared in Figure 16. The desired value is the force in the fixed coordinate system for the actual rotation angle of the two axes when there is only ideal longitudinal propulsion. In other words, the desired forces are the decomposition force to the rotation angle measured by the encoder values of the two servo motors, based on the propulsion model of the fixed thruster. These are ideal values that do not consider force attenuation or tangential force due to the rotation of the thruster. In the initial position, the propulsion occurs in only the X-axis direction, but as the two axes rotate simultaneously, the direction of the propulsion changes. For each force and moment, the steady-state error and root mean square error (RMSE) are defined as performance indices and are shown in Table 6.

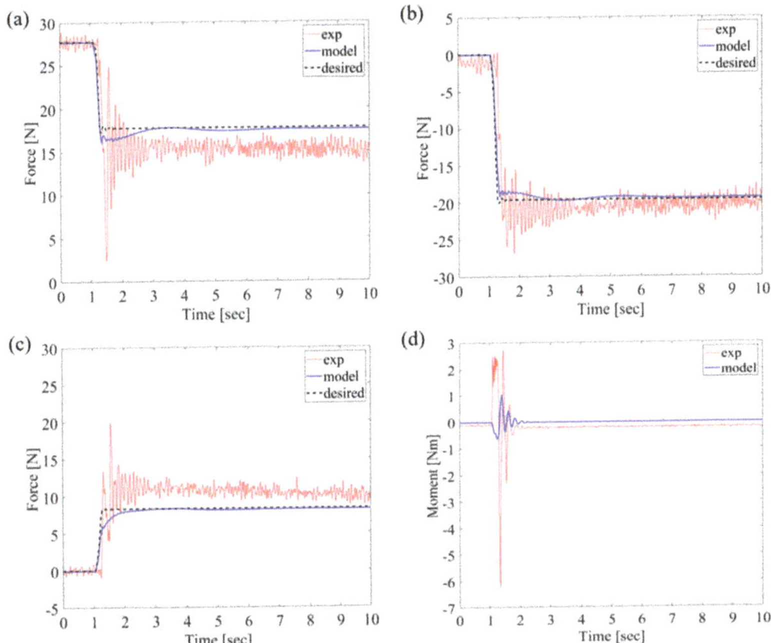

Figure 16. Comparison between experimental results and model under the rotational input with 65% PWM: (**a**) $f_X(t)$, (**b**) $f_Y(t)$, (**c**) $f_Z(t)$ and (**d**) $M_Z(t)$.

Table 6. Steady-state error and root mean square error (RMSE).

Performance Index	f_X	f_Y	f_Z	M_Z
Steady-state error	2.29 N	0.84 N	2.07 N	0.20 N·m
RMSE	2.66 N	2.20 N	2.65 N	0.63 N·m

The simulation results from the derived model followed the waveform changes in the experimental results well. The change in force on the rotation of the two axes was modeled, but the model did not include high-frequency oscillations from the rotation of the propeller. As the motor encoders measuring the rotation angle are very precise, the steady-state error observed in the X- and Z-axial forces can be considered as interference from forces from the water surface and water tanks. Overshoots in the experimental values are caused by the vibration of the gimbal structure. It was observed that more vibrations caused by step inputs occurred than in the modeling experiments.

6. Conclusions

This study showed an empirical model of the force and moment of a 2-DoF azimuth thruster. A signal compression method was used to extract frequency response characteristics within mechanical limits. The longitudinal thrust and lateral thrust and reaction moments generated by the rotation of each axis were measured through repeated experiments, and the model was derived by curve fitting to the Bode plot obtained from the experimental results. We validated the derived model by comparing the simulation values with the experimental results on the step inputs of the two axes. From the comparison results, the causes of modeling errors were analyzed. This study shows the trend of change in coefficients of the azimuth thrust model for thruster force, and normalization for the sensitivity analysis of model coefficients will be studied in the future.

The experimental results showed that the gimbal structure has a significant impact on the force and moment of the thruster. The difference between the first-axis rotation model and the second-axis rotation model was also due to the gimbal structure. A design that minimizes the gimbal is needed when fabricating an azimuth thruster. The modeling procedure presented in this study could be applied to develop vectoring thrusters that control the direction of the propellant. It could also help to design controllers for underwater robots with azimuth thrusters.

Author Contributions: Conceptualization, S.J.; methodology, S.J.; software, C.-S.J.; validation, S.J., C.-S.J. and G.K.; formal analysis, G.K.; data curation, G.K.; writing—original draft preparation, C.-S.J.; writing—review and editing, S.J.; visualization, C.-S.J.; supervision, S.J.; project administration, S.J.; funding acquisition, I.L. All authors have read and agreed to the published version of the manuscript.

Funding: This work was supported by the National Research Foundation of Korea (NRF) grant funded by the Korean government (MSIP) through GCRC-SOP (No.2011-0030013).

Institutional Review Board Statement: Not applicable.

Informed Consent Statement: Not applicable.

Data Availability Statement: The data presented in this study are available on request from the corresponding author.

Conflicts of Interest: The authors declare no conflict of interest.

References

1. Yoerger, D.R.; Cooke, J.G.; Slotine, J.-J.E. The influence of thruster dynamics on underwater vehicle behavior and their incorporation into control system design. *IEEE J. Ocean. Eng.* **1990**, *15*, 167–178. [CrossRef]
2. Healey, A.J.; Rock, S.M.; Cody, S.; Miles, D.; Brown, J.P. Toward an improved understanding of thruster dynamics for underwater vehicles. *IEEE J. Ocean. Eng.* **1995**, *20*, 354–361. [CrossRef]

3.   Bachmayer, R.; Whitcomb, L.L.; Grosenbaugh, M.A. An accurate four-quadrant nonlinear dynamical model for marine thrusters: Theory and experimental validation. *IEEE J. Ocean. Eng.* **2000**, *25*, 146–159. [CrossRef]

4.   Blanke, M.; Lindegaard, K.P.; Fossen, T.I. Dynamic model for thrust generation of marine propellers. *IFAC Proc. Vol.* **2000**, *33*, 353–358. [CrossRef]

5.   Kim, J.; Han, J.; Chung, W.K.; Yuh, J.; Lee, P.M. Accurate and practical thruster modeling for underwater vehicle. In Proceedings of the IEEE International Conference on Robotics and Automation (ICRA), Barcelona, Spain, 18–22 April 2005; pp. 175–180. [CrossRef]

6.   Kim, J.; Han, J.; Chung, W.K.; Yuh, J. Accurate thruster modeling with non-parallel ambient flow for underwater vehicle. In Proceedings of the IEEE/RSJ International Conference on Intelligent Robots and Systems (IROS), Edmonton, AB, Canada, 2–6 August 2005; pp. 978–983. [CrossRef]

7.   Kim, J.; Chung, W.K. Accurate and practical thruster modeling for underwater vehicles. *Ocean Eng.* **2006**, *33*, 566–586. [CrossRef]

8.   Yu, L.; Greve, M.; Druckenbrod, M.; Abdel-Maksoud, M. Numerical analysis of ducted propeller performance under open water test condition. *J. Mar. Sci. Technol.* **2013**, *18*, 381–394. [CrossRef]

9.   Sparenberg, J.A. On the axial force induced on a duct by a propulsor. *J. Mar. Sci. Technol.* **1997**, *2*, 53–61. [CrossRef]

10.   Song, Y.S.; Arshad, M.R. Thruster modeling for a Hovering Autonomous Underwater Vehicle considering thruster-thruster and thruster-hull interaction. In Proceedings of the 2016 IEEE International Conference on Underwater System Technology: Theory and Applications (USYS), Penang, Malaysia, 13–14 December 2016. [CrossRef]

11.   Boehm, J.; Berkenpas, E.; Henning, B.; Rodriguez, M.; Shepard, C.; Turchik, A. Characterization, Modeling, and Simulation of an ROV Thruster using a Six Degree-of-Freedom Load Cell. In Proceedings of the OCEANS 2018 MTS/IEEE, Charleston, SC, USA, 22–25 October 2018. [CrossRef]

12.   Odetti, A.; Altosole, M.; Bruzzone, G.; Caccia, M.; Viviani, M. Design and construction of a modular pump-jet thruster for autonomous surface vehicle operations in extremely shallow water. *J. Mar. Sci. Technol.* **2019**, *7*, 222. [CrossRef]

13.   Odetti, A.; Altosole, M.; Bibuli, M.; Bruzzone, G.; Caccia, M.; Viviani, M. Advance Speed-Hull-Pump-Jet Interactions in Small ASV. *Prog. Mar. Sci. Technol.* **2020**, *5*, 197–206. [CrossRef]

14.   Fagundes Gasparoto, H.; Chocron, O.; Benbouzid, M.; Siqueira Meirelles, P. Advances in Reconfigurable Vectorial Thrusters for Adaptive Underwater Robots. *J. Mar. Sci. Technol.* **2021**, *9*, 170. [CrossRef]

15.   Jin, S.; Kim, J.; Kim, J.; Seo, T.W. Six-degree-of-freedom hovering control of an underwater robotic platform with four tilting thrusters via selective switching control. *IEEE/ASME Trans. Mechatron.* **2015**, *20*, 2370–2378. [CrossRef]

16.   Liang, C.C.; Cheng, W.H. The optimum control of thruster system for dynamically positioned vessels. *Ocean Eng.* **2004**, *31*, 97–110. [CrossRef]

17.   Amini, H.; Sileo, L.; Steen, S. Numerical calculation of propeller shaft loads on azimuth propulsors in oblique inflow. *J. Mar. Sci. Technol.* **2012**, *17*, 403–421. [CrossRef]

18.   Berchiche, N.; Krasilnikov, V.; Koushan, K. Numerical Analysis of Azimuth Propulsor Performance in Seaways: Influence of Oblique Inflow and Free Surface. *J. Mar. Sci. Technol.* **2018**, *6*, 37. [CrossRef]

19.   Jin, S.; Kim, J.; Lee, S.; Kim, J.; Seo, T.W. Empirical modeling of rotating thruster for underwater robotic platform. *J. Mar. Sci. Technol.* **2015**, *20*, 118–126. [CrossRef]

20.   Jin, S.; Bak, J.; Kim, J.; Seo, T.; Kim, H.S. Switching PD-based sliding mode control for hovering of a tilting-thruster underwater robot. *PLoS ONE* **2018**, *13*, e0194427. [CrossRef] [PubMed]

21.   Kim, H.; Sitti, M.; Seo, T. Tail-Assisted Mobility and Stability Enhancement in Yaw and Pitch Motions of a Water-Running Robot. *IEEE/ASME Trans. Mechatron.* **2017**, *22*, 1207–1217. [CrossRef]

22.   Lee, M.C.; Aoshima, N. Identification and its evaluation of the system with a nonlinear element by signal compression method. *Trans. the Soc. Instrum. Control Eng.* **1989**, *25*, 729–736. [CrossRef]

23.   Yang, S.Y.; Lee, M.C.; Lee, M.H.; Arimoto, S. Measuring System for Development of Stroke-Sensing Cylinder for Automatic Excavator. *IEEE Trans. Ind. Electron.* **1998**, *45*, 376–384. [CrossRef]

24.   Kallu, K.D.; Abbasi, S.J.; Hamza, K.H.; Wang, J.; Lee, M.C. Implementation of a TSMCSPO Controller on a 3-DOF Hydraulic Manipulator for Position Tracking and Sensor-Less Force Estimation. *IEEE Access* **2019**, *7*, 177035–177047. [CrossRef]

25.   Ogata, K. Chapter 7 Control Systems Analysis and Design by the Frequency-Response Method. In *Modern Control Engineering*, 4th ed.; Prentice-Hall: Hoboken, NJ, USA, 2001; pp. 398–521.

Article

Energy-Efficient Hip Joint Offsets in Humanoid Robot via Taguchi Method and Bio-inspired Analysis

Jihun Kim [1], Jaeha Yang [1], Seung Tae Yang [1], Yonghwan Oh [2] and Giuk Lee [1,*]

[1] School of Mechanical Engineering, Chung-Ang University, Seoul 06974, Korea; jeyjey6@naver.com (J.K.); nca010552@naver.com (J.Y.); hilton99@cau.ac.kr (S.T.Y.)

[2] Korea Institute of Science and Technology, Seoul 02792, Korea; oyh@kist.re.kr

* Correspondence: giuklee@cau.ac.kr

Received: 26 August 2020; Accepted: 15 October 2020; Published: 18 October 2020

Abstract: Although previous research has improved the energy efficiency of humanoid robots to increase mobility, no study has considered the offset between hip joints to this end. Here, we optimized the offsets of hip joints in humanoid robots via the Taguchi method to maximize energy efficiency. During optimization, the offsets between hip joints were selected as control factors, and the sum of the root-mean-square power consumption from three actuated hip joints was set as the objective function. We analyzed the power consumption of a humanoid robot model implemented in physics simulation software. As the Taguchi method was originally devised for robust optimization, we selected turning, forward, backward, and sideways walking motions as noise factors. Through two optimization stages, we obtained near-optimal results for the humanoid hip joint offsets. We validated the results by comparing the root-mean-square (RMS) power consumption of the original and optimized humanoid models, finding that the RMS power consumption was reduced by more than 25% in the target motions. We explored the reason for the reduction of power consumption through bio-inspired analysis from human gait mechanics. As the distance between the left and right hip joints in the frontal plane became narrower, the amplitude of the sway motion of the upper body was reduced. We found that the reduced sway motion of the upper body of the optimized joint configuration was effective in improving energy efficiency, similar to the influence of the pathway of the body's center of gravity (COG) on human walking efficiency.

Keywords: humanoid robot; energy efficiency; Taguchi method

---

1. Introduction

As quality of life improves, the demand for humanoid robots to help or replace humans in various activities is increasing. However, the mobility of humanoid robots is limited by their main power source, batteries, which have limited available power. To improve the performance of humanoid robots, energy efficiency must be increased to provide longer operation periods or enable the use of lighter and smaller batteries.

Several studies have aimed to improve energy efficiency by reducing power consumption in the actuators of humanoid robots. Zorjan et al. [1] observed that the rotation of the hip joint from its original orientation improves energy efficiency by reducing and distributing the joint's mechanical power in the hip joints. Lee et al. [2] proposed a biarticular design for humanoid robot legs using a redundantly actuated parallel mechanism to improve energy efficiency. Negrello et al. [3] designed a four-bar linkage for knee and ankle joints, thus decreasing the mass and moment of inertia of the lower body to enhance energy efficiency. Tsagarakis et al. [4] studied an asymmetric compliant antagonistic joint design to improve performance mobility, which is closely related to energy efficiency.

Although previous research has improved the energy efficiency of humanoid robots, no study has considered the offset between hip joints to this end. This offset is the gap between adjacent joints along a coordinate axis (e.g., a roll–yaw–pitch hip joint, where the offset is represented by each gap between the roll and yaw joint and between the yaw and pitch joint along the X, Y, and Z axes). Mechanically, it is difficult to overlap multiple joints at a single intersection point and, therefore, the effect of joint offset on the humanoid robot should be investigated.

In this study, we optimized the hip joint offset in a humanoid robot to improve energy efficiency by reducing consumption at the actuating joints. Optimization considered bipedal locomotion, which is representative of humanoid robots, by including turning, forward, backward, and sideways walking motions. We analyzed the mechanical power consumption of the actuated hip joints using a humanoid model, implemented in physics simulation software (MuJoCo, Roboti, Seattle, DC, USA) for energy optimization. Specifically, the objective function was the root-mean-square (RMS) mechanical power consumption from the actuated hip joints.

The complex dynamics of multi-joint humanoid robot legs hinder the ability of investigations of the joint offset effect to obtain, for instance, an analytical solution. Therefore, it is more appropriate to use numerical optimization for this study. However, it becomes impractical, given the time-consuming generation of many complex simulation models of humanoid robot legs with different joint offset configurations.

To minimize the time-consuming efforts of numerical optimization, we implemented the Taguchi method, which is an experimental approach for the robust, optimal design of products under various conditions. As the method conducts optimization based on an orthogonal array, it can minimize the number of simulation models required, thus reducing the optimization execution time. For the Taguchi method, we set the joint offsets as control factors and the four abovementioned types of motions as noise factors.

2. Implementation of the Taguchi Method

The Taguchi method is an experimental optimization method, aiming to improve the quality of products in various conditions [5–8]. In addition, the experimental volume of the Taguchi method is minimized by designing experiments with an orthogonal array, unlike traditional factorial design. In this study, the Taguchi method enabled a robust hip joint offset design that can stabilize the performance of humanoid robots under variable motions, with a lower number of trials than other numerical optimization methods.

The Taguchi method considers two types of variables, namely control and noise factors. Experiments for each factor were designed by applying the corresponding orthogonal array, with the array for control factors being the inner array and that for noise factors being the outer array. The inner and outer array combination constituted the crossed array, which established orthogonality among the individual arrays. Each inner array sample was tested for a full set of experiments from the outer array. Hence, the orthogonal array allowed the exploration of the most influential factors with a small experiment volume. To evaluate optimization, we adopted the signal-to-noise ratio (SNR), whose calculation depends on the objective function value.

2.1. Objective Function

We defined the objective function as the sum of RMS power consumption from the right hip joint:

$$Pow_{rms} = R_{rms} + Y_{rms} + P_{rms} = \sqrt{\frac{\sum_{i=1}^{n_T} R_i^2}{n_T}} + \sqrt{\frac{\sum_{i=1}^{n_T} Y_i^2}{n_T}} + \sqrt{\frac{\sum_{i=1}^{n_T} P_i^2}{n_T}} \tag{1}$$

where Pow_{rms} is the sum of the RMS power consumption and R_{rms}, Y_{rms}, and P_{rms} represent the RMS power consumption at the roll, yaw, and pitch joints, respectively. R_i, Y_i, and P_i represent the power consumption of the ith sample at the roll, yaw, and pitch joints, respectively, and n_T is the number of samples until time T.

2.2. Control and Noise Factors

In the Taguchi method, control factors are design parameters that can be adjusted during optimization. Noise factors can have several meanings, but they typically refer to the operating conditions of the target system. Hence, both control and noise factors influence the value of the objective function. We defined the joint offsets within the hip as control factors and the humanoid motion types as noise factors. More details about these parameters are provided in Section 3.

2.3. SNR Calculations

The criteria for the Taguchi method are classified as smaller-the-better, larger-the-better, and nominal-the-best, according to their characteristics. We used the smaller-the-better variant, given the objective of minimizing power consumption at the humanoid hip. According to the smaller-the-better definition, we calculated the SNR for the Taguchi method using the objective function value of each trial:

$$\text{SNR} = -10\,log_{10}\left(\frac{Pow_{rms\_for}^2 + Pow_{rms\_back}^2 + Pow_{rms\_side}^2 + Pow_{rms\_turn}^2}{4}\right), \tag{2}$$

where Pow_{rms_for}, Pow_{rms_back}, Pow_{rms_side}, and Pow_{rms_turn} are the calculated Pow_{rms} values from the forward, backward, and sideways walking, as well as turning motions. In the smaller-the-better case, the higher SNR represents better results for optimization.

2.4. Orthogonal Array of Experiment Set

We prepared two experimental sets for optimization. The orthogonal array of each set was selected according to the number and level of control factors. The first experimental set consisted of an L_{18} ($2^1 \times 3^7$) orthogonal array for eight control factors. The second experimental set did not need to include the optimized control factors provided by the first experimental set. Thus, six control factors with small SNR variations were omitted from the second experimental set, which consisted of an L_9 (3^3) orthogonal array for the three remaining control factors.

2.5. Detailed Procedure of the Taguchi Method

Since the Taguchi Method is not a commonly implemented optimization method, we summarized the detailed procedure for performing the Taguchi Method for better understanding:
1. Set the objective function and design parameters for optimization;
2. Classify the design parameters into control factors and noise factors;
3. Select the SNR calculation formula based on the characteristics of the objective function;
4. Determine the design levels for the control factors;
5. Select an appropriate orthogonal array for the experiment set;
6. Perform experiments according to the experiment set;
7. Analyze experimental data and derive optimized results using the SNR calculation formula;
8. If the optimized result is not sufficient, design a new experiment set for additional experiments;
9. Eliminate less influential control factors in additional experiments to simplify the experiment sets.

3. Experimental Design

3.1. Humanoid Kinematics

The humanoid robot used for simulation was designed to have a similar size to a young man. We selected the length and mass of each segment and the size of the humanoid robot by referring to the average body size of Koreans aged 16 to 19 [9]. Other kinematic parameters for the detailed design of the humanoid robot were based on the MAHRU humanoid robot, previously developed in our institute [10]. The robot had 22 degrees of freedom, weighed 57 kg, and had a height of 1.216 m and shoulder width of 0.5 m. The humanoid robot structure is detailed in Figure 1. Its thigh length and weight were 0.4 m and 5.5 kg, respectively, whereas its calf length and weight were 0.3 m and 3.5 kg, respectively, and its ankle length and weight were 0.07 m and 0.25 kg, respectively. Each foot was 0.026 m in height, 0.25 m in length, 0.1 m in width, and 0.75 kg in weight.

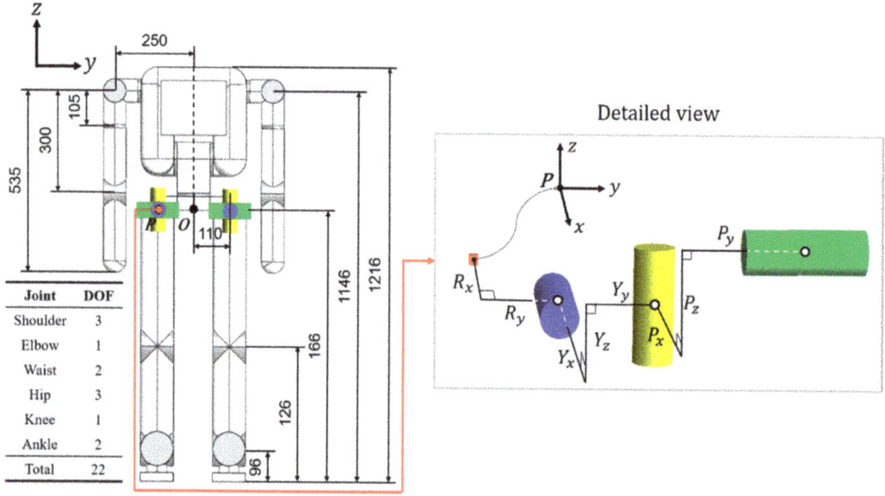

Figure 1. Humanoid robot simulation model (units in millimeters) for optimization. The parameters in the detailed view correspond to joint offsets. The design parameters were configured based on the front view of the hip joint for the right leg. The roll, yaw, and pitch joints are represented as blue, yellow, and green colors, respectively.

Although the original width between the left and right foot was 0.22 m, we varied it in each test model, depending on the selected offset in the hip joint along the Y axis. For example, a test model with a 0.03 m offset was designed with a width of 0.28 m between the feet.

3.2. Control Factors: Joint Offsets

We set the offsets among the three joints (roll, yaw, and pitch) in the humanoid robot hip as control factors. The hip joint used for simulation was composed in the sequence of roll, yaw, and pitch joints. The offset between the torso and roll joints along the X and Y axes was selected as a parameter because this joint did not have a preceding offset. The offset between the torso and roll joints along the Z axis was fixed and disregarded as a parameter because it changed the humanoid leg length.

For the roll–yaw joint and yaw–pitch joint, all offsets along the X, Y, and Z axes were selected. The offsets along the Z axis were negative because it was challenging to place a following joint higher than the preceding joint in practical mechanisms. The offsets at the joints that serve as parameters are detailed in Figure 1 (right graph), with each offset being denoted by the letter of the following joint and offset direction. For example, the offset between the torso and roll joints along the Y axis is denoted by R_y, and the offset between the yaw and pitch joints along the Z axis is denoted by P_z.

3.3. Noise Factors: Humanoid Motions

We set turning, forward, backward, and sideways walking motions as noise factors. As forward and backward are the most basic humanoid robot motions, they were selected as target motions. Moreover, turning and sideways walking were selected, given their distinctiveness compared with forward and backward motions, despite their lower occurrence frequencies. The walking patterns were generated by using the zero-moment point (ZMP) controller designed for the MAHRU humanoid robot [10].

For fair comparison during the optimization process, we set the movement pattern of each model's feet to be the same. We fixed the following parameters related to feet motion generation as constant values:

- Step length: 200 mm for forward and backward walking and 100 mm for sideways walking;
- Maximum foot clearance (maximum lifting height of the feet during gait): 80 mm;
- Single and double support time for a stride: 800 ms and 200 ms;

The amplitude of the center-of-mass for generating the ZMP trajectory was set as a constant value of 650 mm.

3.3.1. Forward and Backward Walking

Forward and backward walking were set identically in the experiments because they occurred along the same axis. The humanoid only moved backward and forward while these motions were executed. During walking, the left foot moved first, and each foot completed three steps in the forward or backward direction. Hence, the humanoid robot moved 1 m in either direction, as shown in Figure 2.

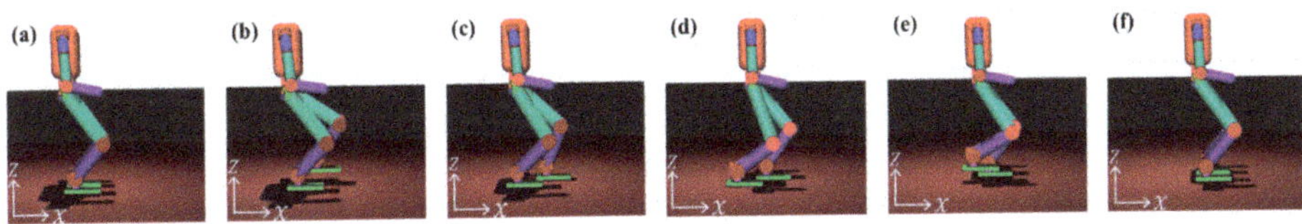

Figure 2. Forward walking. (**a**) Initial position, (**b**) left foot lifting, (**c**) left foot lowering after swing, (**d**) center-of-mass forward displacement, (**e**) right foot lifting, and (**f**) motion completion.

3.3.2. Sideways Walking

Sideways walking is a perpendicular motion, with respect to forward and backward walking, without rotating the humanoid robot body. Starting with the left foot, each one completed three steps. Thus, the humanoid robot moved 300 mm to the left after sideways walking, which proceeded as shown in Figure 3.

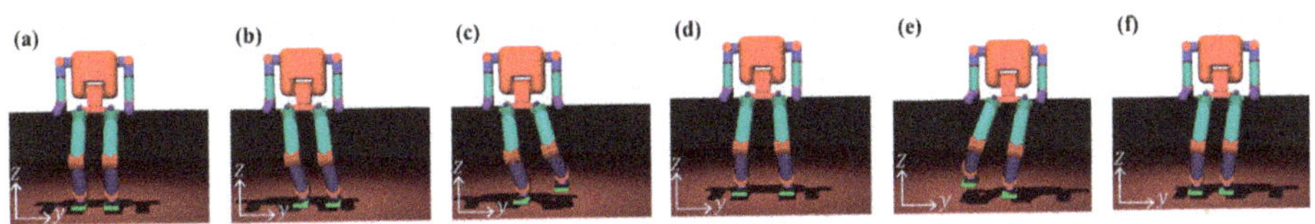

Figure 3. Sideways walking. (**a**) Initial position, (**b**) moving zero-moment point above the right foot, (**c**) left foot lifting, (**d**) left foot lowering after swing, (**e**) moving zero-moment point above the left foot and right foot lifting, (**f**) right foot lowering and motion completion.

3.3.3. Turning

Unlike the previous motions, turning changes only the orientation of the body at a fixed position. The turning angle was set to 20° for the first experimental set and 15° for the second one, being smaller because the corresponding test model for the second stage had a smaller supporting polygon, leading to instability at larger angles. The supporting polygon is further explained in the corresponding discussion in Section 6. Starting with the left foot, the humanoid turned three times per trial, reaching overall turning angles of 60° and 45° for the first and second experimental sets, respectively. Figure 4 illustrates the turning motion.

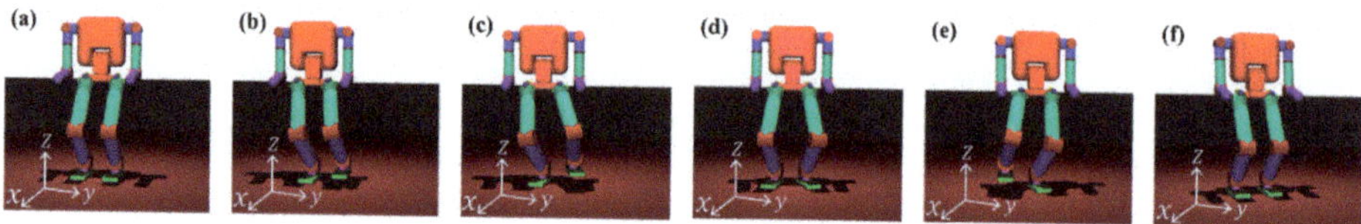

Figure 4. Turning. (**a**) Initial position, (**b**) moving zero-moment point above the right foot, (**c**) left-foot lifting and both left leg and body turning, (**d**) left foot lowering, (**e**) right foot lifting and both right leg and body turning, and (**f**) motion completion.

4. Hip Joint Offset Optimization

4.1. First Experimental Set

The first experimental set consisted of 18 trials, according to the L_{18} ($2^1 \times 3^7$) orthogonal array. Tables 1 and 2 list the parameter levels and results per trial according to these levels, respectively. As shown in Table 2, the SNR of the first test model was the highest at −23.923 when all control factor levels were 1. In contrast, the SNR of the third test model was the lowest. A higher SNR indicates that the objective function reached closer to the optimal value. Figure 5 shows the SNR of each factor according to its level. The highest SNR was achieved at the level of $P_z = 2$, $R_x = 2$, $R_y = 1$, $Y_x = 3$, $Y_y = 1$, $Y_z = 2$, $P_x = 1$, and $P_y = 1$, and the factors with high sensitivities were R_y, Y_y, and P_y, with respective SNR variations of 1.90, 1.05, and 0.74. Factors with SNR variations below 0.3 were considered as insensitive.

Table 1. Levels of control factors for the first experimental set.

Level	Control Factor (mm)							
	P_z	R_x	R_y	Y_x	Y_y	Y_z	P_x	P_y
1	−10	−10	−10	−10	−10	−20	−10	−10
2	0	0	0	0	0	−10	0	0
3	–	10	10	10	10	0	10	10

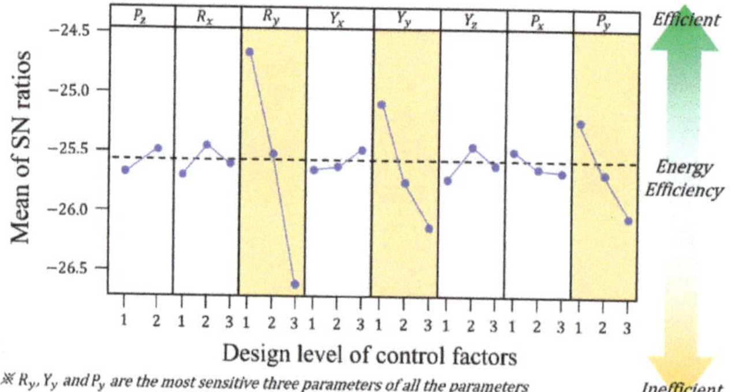

Figure 5. Signal-to-noise ratios (SNRs) of the eight parameters during the first experiment. The three most influential factors are highlighted in yellow.

Table 2. Design, results, and SNRs of first experimental set. (The highest SNR is boldfaced.)

Test	Control Factor								Noise Factor				SNR
	P_z	R_x	R_y	Y_x	Y_y	Y_z	P_x	P_y	Forward Walking	Backward Walking	Sideways Walking	Turning	
									Objective Function (W)				
1	1	1	1	1	1	1	1	1	17.237	19.384	13.928	10.970	**−23.923**
2	1	1	2	2	2	2	2	2	21.852	21.427	19.928	14.353	−25.854
3	1	1	3	3	3	3	3	3	24.439	25.420	26.282	17.854	−27.507
4	1	2	1	1	2	2	3	3	19.349	20.095	18.557	13.403	−25.126
5	1	2	2	2	3	3	1	1	20.484	21.175	19.856	13.993	−25.617
6	1	2	3	3	1	1	2	2	21.263	22.027	21.309	14.951	−26.061
7	1	3	1	2	1	3	2	3	18.467	19.167	17.206	12.542	−24.631
8	1	3	2	3	2	1	3	1	21.706	20.175	18.439	13.416	−25.435
9	1	3	3	1	3	2	1	2	23.512	24.215	24.664	16.653	−27.043
10	2	1	1	3	3	2	2	1	18.416	19.150	16.947	12.493	−24.583
11	2	1	2	1	1	3	3	2	19.613	20.351	18.673	13.459	−25.214
12	2	1	3	2	2	1	1	3	23.605	24.906	24.626	16.663	−27.121
13	2	2	1	2	3	1	3	2	19.379	20.073	18.582	13.406	−25.130
14	2	2	2	3	1	2	1	3	18.022	19.990	18.280	11.949	−24.776
15	2	2	3	1	2	3	2	1	21.555	22.367	21.442	15.045	−26.157
16	2	3	1	3	2	3	1	2	18.453	19.177	17.186	12.513	−24.625
17	2	3	2	1	3	1	2	3	22.146	23.095	22.925	15.855	−26.534
18	2	3	3	2	1	2	3	1	20.096	21.019	19.397	14.175	−25.511

4.2. Second Experimental Set

As the optimal points do not appear in the SNR graph of Figure 5, we considered only the three highly sensitive factors (R_y, Y_y, and P_y) for the second experimental set. The kinematics of the humanoid model for the second experimental set were designed using the optimal result from the first experimental set. Specifically, the value of each factor, including insensitive ones, was set to that retrieving the highest SNR in the first experimental set.

The second experimental set consisted of nine trials, according to the L_9 (3^3) orthogonal array, with three factors. Tables 3 and 4 list the parameter levels and results per trial according to these levels, respectively. The SNR was calculated as in the first experiment, obtaining the highest SNR at the ninth trial of –22.614 for the level of $R_y = 3$, $Y_y = 3$, and $P_y = 2$.

Table 3. Levels of control factors for the second experimental set.

Level.	Control Factor (mm)		
	R_y	Y_y	P_y
1	0	0	0
2	−2.5	−2.5	−2.5
3	−5	−5	−5

Table 4. Design, results, and SNRs of the second experimental set. (The highest SNR is boldfaced.)

Test	Control Factor			Noise Factors				SNR
	R_y	Y_y	P_y	Forward Walking	Backward Walking	Sideways Walking	Turning	
				Objective Function (W)				
1	1	1	1	16.591	17.387	13.951	10.832	−23.471
2	1	2	2	16.212	16.953	13.572	10.518	−23.248
3	1	3	3	15.721	16.520	13.111	10.198	−22.988
4	2	1	2	16.016	16.747	13.347	10.352	−23.130
5	2	2	3	15.528	16.322	12.875	10.040	−22.867
6	2	3	1	15.764	16.532	13.120	10.218	−23.001
7	3	1	3	15.349	16.122	12.660	9.881	−22.750
8	3	2	1	15.573	16.337	12.901	10.034	−22.881
9	3	3	2	15.113	15.871	12.491	9.690	**−22.614**

The SNRs, according to the level of each control factor in the second experimental set, are shown in Figure 6. The highest SNR was obtained when the three factors were at level 3, with R_y being the most sensitive factor with an SNR variation of 0.49 across all levels. The SNR variation for both Y_y and P_y was 0.25, with their sensitivities being lower than that of R_y. Although the results did not reach the optimal solution, we ended the optimization because of the limited range of feasible joint offsets.

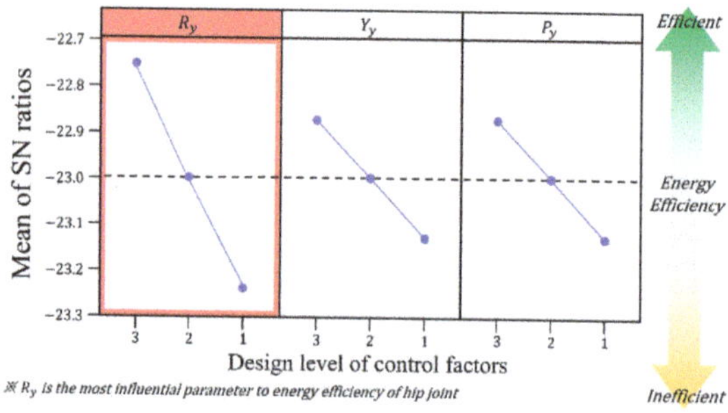

Figure 6. SNRs of three sensitive parameters during the second experiment. The most influential factor is R_y, highlighted in red.

5. Validation of Optimal Hip Joint Offsets

The near-optimal offsets of hip joints that minimize power consumption were obtained from the two experimental sets. Using the optimized model, we verified that power consumption was indeed reduced for various humanoid robot motions. The optimal values of the parameters are listed in Table 5, and Figure 7 shows the joint configuration and structure of the model after optimization in the second experimental set.

Table 5. Parameter values of the model after the second optimization.

Parameter	P_z	R_x	R_y	Y_x	Y_y	Y_z	P_x	P_y
Value (mm)	0	0	−15	10	−15	0	−10	−15

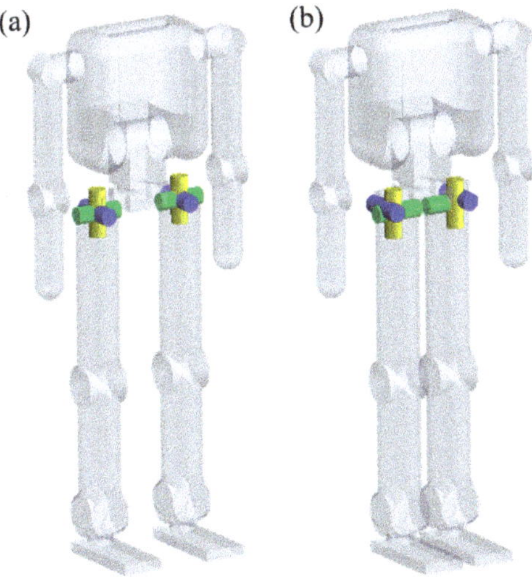

Figure 7. Hip joint configuration of (**a**) the original model and (**b**) the model after the second optimization.

5.1. Forward and Backward Walking

Table 6 lists the power consumption in the hip, using the original model and the optimized models after the first and second experiments for forward and backward walking. For forward walking, the RMS power at the hip joint reduced by 18.07% and 26.92% for the models after the first and second optimizations, respectively, compared with the original model. For backward walking, the corresponding reductions were 17.26% and 25.68%.

Table 6. Root-mean-square (RMS) power reduction using an optimized model in humanoid hip joints during target motions.

Model	Forward Walking		Backward Walking		Sideways Walking		Turning	
	Pow_{rms} (W)	Reduction (%)	Pow_{rms} (W)	Reduction (%)	Pow_{rms} (W)	Reduction (%)	Pow_{rms} (W)	Reduction (%)
Original	20.250	–	21.013	–	19.868	–	15.325	–
First Optimization	16.591	18.07	17.387	17.26	13.951	29.78	11.983	21.81
Second Optimization	14.799	26.92	15.617	25.68	12.365	37.76	10.713	30.09

5.2. Sideways Walking

Table 6 lists the power consumption in the hip joint for sideways walking. Compared to the original model, consumption decreased by 29.78% and 37.76% for the models after the first and second optimization, respectively. Like for forward and backward walking, power consumption gradually reduced from the original model to the models after the first and second optimizations.

5.3. Turning

The power consumption for turning using the original and optimized models is listed in Table 6. Note that the rotation angle for turning in the humanoid models was set to 15° equally. Like in the previous motions, consumption reduced in the hip joint, in this case by 21.81% and 30.09% for the models after the first and second optimization, respectively.

6. Discussion

We investigated the optimization for reducing power consumption at the hip of a humanoid robot via the Taguchi method. The two-stage optimization achieved a larger reduction in consumption than that obtained using only one stage. Power consumption at the hip was mainly influenced by the Y axis offset. Therefore, the closer the sides of the hip joint, the lower the energy consumption. The sum of RMS power consumption from the right hip joint (Pow_{rms}) over time during the motions is shown in Figure A1 and supplementary materials.

Humanoid robots implement zero-moment point walking to keep the point position inside the supporting polygon and thus stabilize the robot's walking. The supporting polygon is a convex hull containing the contact points between the feet and ground. If both feet are close due to the hip joints being close, the supporting polygon becomes narrower. During single-leg support, the robot sways its upper body from side to side to place the zero-moment point inside the supporting polygon. Consequently, placing the hip joints closer reduces the area to move the zero-moment point to keep it within the supporting polygon while switching from double- to single-leg support. Then, the moment acting on the roll joint is decreased by the smaller moment arm of the upper body, related to the roll joint. Therefore, joint offsets along the Y axis are advantageous to improving energy efficiency by reducing power consumption in the roll joint, as both hip joints are close.

We explored the reason for the reduction of power consumption through bio-inspired analysis from human gait mechanics. The finding in this study is similar to the influence of the body's center of gravity (COG) pathway on human walking efficiency. Minimizing the body's COG, displaced from a level line of progression, is considered to be a major mechanism for reducing the muscular effort of walking and, consequently, for saving energy. The least energy would be used if the weight being carried remained at a constant height and followed a single central path. No additional lifting effort would then be needed to recover from the intermittent falls downward or laterally [11].

As an additional analysis, we compared the maximum required torque at the joint after optimization. The reduction of actuating torque is an important issue for reducing the weight of the actuators. Except for the pitch joint during the turning motion, all maximum torques generated at each joint were reduced by a minimum of 0.95% and a maximum of 44.67%. Before the optimization, the maximum torque of all joints was generated at a value of 70.91 Nm at the roll joint during the sideways walking motion. Through the optimization, this value was reduced to 59.15 Nm, which was also the maximum torque for the optimized model.

Although we obtained valuable insights on offset design, some limitations of this study remain to be addressed. We considered only mechanical power to represent power consumption, because it is impossible to calculate the electrical power dissipated in the actuator before selecting the motor and other electronic components. In addition, although the negative mechanical power can be restored using a regeneration system, we neglected this aspect because humanoid robots typically do not implement a regeneration system.

We expect to use the findings in this study to design an improved humanoid robot. By improving energy efficiency, the operating time of humanoid robots can be extended. As for future work, we plan to verify energy efficiency optimization while a humanoid robot performs various tasks. Moreover, we intend to extend joint offset optimization to include knee and ankle joints.

Supplementary Materials: The following are available online at http://www.mdpi.com/2076-3417/10/20/7287/s1. Video S1: The sum of RMS power consumption from the right hip joint of the original, first, and second optimized humanoid robot at the hip joint during the target motions.

Author Contributions: Conceptualization, J.K., J.Y., S.T.Y., Y.O., and G.L.; methodology, J.K., J.Y., and G.L; software, J.K., J.Y., Y.O., and G.L.; validation, J.K., J.Y., and G.L.; data analysis, J.K., J.Y., and G.L.; writing—original draft preparation, J.K. and G.L.; writing—review and editing, J.K., J.Y., Y.O., and G.L.; visualization, J.K. and G.L.; supervision, Y.O. and G.L.; project administration, Y.O. and G.L.; funding acquisition, Y.O. and G.L. All authors have read and agreed to the published version of the manuscript.

Funding: This research was funded by the Korea Institute of Science and Technology, KIST, Institutional Program (Project No. 2E29460-19-018) and the Chung-Ang University Research Scholarship Grants in 2020.

Acknowledgments: The authors thank Seung Jae Yoo from the Korea Institute of Science and Technology for his contributions to this work.

Conflicts of Interest: The authors declare no conflict of interest.

Appendix A

Figure A1 shows the sum of RMS power consumption from the right hip joint (Pow_{rms}) over time of the original, first, and second humanoid robot at the hip joint during the target motions.

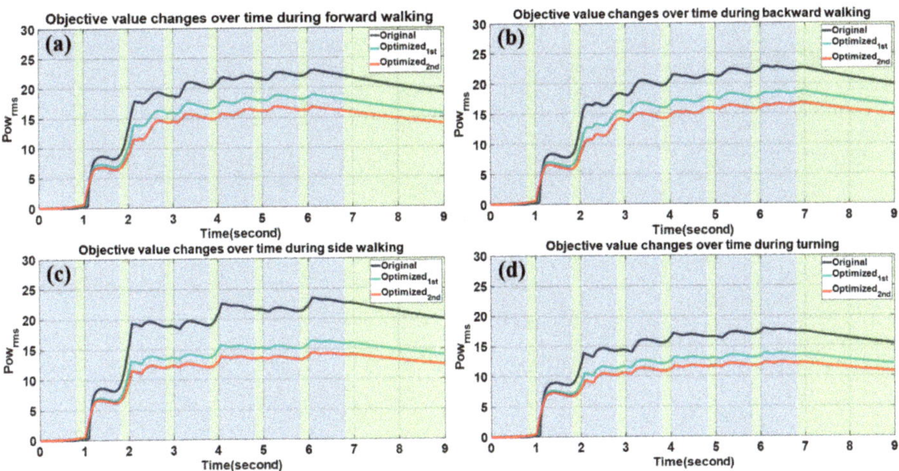

Figure A1. Objective value changes over time during (**a**) forward walking, (**b**) backward walking, (**c**) sideways walking, and (**d**) turning. Optimized 1st and 2nd indicate the models after the first and second optimization, respectively. Blue and green colored backgrounds indicate a single and double support phase during target motion, respectively.

References

1. Zorjan, M.; Hugel, V.; Blazevic, P.; Borovac, B. Influence of rotation of humanoid hip joint axes on joint power during locomotion. *Adv. Robot.* **2015**, *29*, 707–719. [CrossRef]
2. Lee, J.; Lee, G.; Oh, Y. Energy-efficient robotic leg design using redundantly actuated parallel mechanism. In Proceedings of the IEEE International Conference on Advanced Intelligent Mechatronics (AIM), Munich, Germany, 3–7 July 2017.

3.   Negrello, F.; Garabini, M.; Catalano, M.G.; Kryczka, P.; Choi, W.; Caldwell, D.G.; Bicchi, A.; Tsagarakis, N.G. Walk-man humanoid lower body design optimization for enhanced physical performance. In Proceedings of the IEEE International Conference on Robotics and Automation (ICRA), Stockholm, Sweden, 16–21 May 2016.

4.   Tsagarakis, N.G.; Morfey, S.; Dallali, H.; Medrano-Cerda, G.A.; Caldwell, D.G. An asymmetric compliant antagonistic joint design for high performance mobility. In Proceedings of the IEEE/RSJ International Conference on Intelligent Robots and Systems (IROS), Tokyo, Japan, 3–7 November 2013.

5.   Tsui, K.L. An overview of Taguchi method and newly developed statistical methods for robust design. *IIE Trans.* **1992**, *24*, 44–57. [CrossRef]

6.   Hong, H.S.; Seo, T.; Kim, D.; Kim, S.; Kim, J. Optimal design of hand-carrying rocker-bogie mechanism for stair climbing. *J. Mech. Sci. Technol.* **2013**, *27*, 125–132. [CrossRef]

7.   Sung, T.; Oh, D.; Jin, S.; Seo, T.W.; Kim, J. Optimal design of a micro evaporator with lateral gaps. *Appl. Therm. Eng.* **2009**, *29*, 2921–2926. [CrossRef]

8.   Jugulum, R.; Taguchi, S. *Computer-Based Robust Engineering: Essentials for DFSS*, 1st ed.; ASQ Quality Press: Milwaukee, WI, USA, 2004; p. 95.

9.   Size Korea. Available online: https://sizekorea.kr/page/report/1 (accessed on 4 January 2020).

10.   Kwon, W.; Kim, H.K.; Park, J.K.; Roh, C.H.; Lee, J.; Park, J.; Kim, W.K.; Roh, K. Biped humanoid robot Mahru III. In Proceedings of the IEEE-RAS International Conference on Humanoid Robots, Pittsburgh, PA, USA, 29 November–1 December 2007.

11.   Perry, J.; Burnfield, J.M. Center of Gravity Alignment Modulation. In *Gait Analysis: Normal and Pathological Function*, 2nd ed.; SLACK Incorporated: Thorofare, NJ, USA, 2010; pp. 40–43.

Publisher's Note: MDPI stays neutral with regard to jurisdictional claims in published maps and institutional affiliations.

MDPI

St. Alban-Anlage 66

4052 Basel

Switzerland

Tel. +41 61 683 77 34

Fax +41 61 302 89 18

www.mdpi.com

Applied Sciences Editorial Office

E-mail: applsci@mdpi.com

www.mdpi.com/journal/applsci